Dedication

To my dad for introducing me to the beauty and wonder of both math and nature. To my bunnies, for being adorable.

Acknowledgements

The author is grateful to Chuyue Huang for help proofreading.

Contents

Preface

There is an unreasonable joy from observing small birds going about their bright, oblivious business on a peaceful Saturday morning. One of the great pleasures of bird watching is that once you have some familiarity, it becomes part of your daily life. You are never bored again while walking outside because there are always birds flying about. To me, mathematics is no different. Once you have reached a certain level of experience, you are never bored again – you see it everywhere, and it enriches your life. This book, I hope, will serve as a beginner bird-watcher's guide. Each chapter is a different "bird". I will describe its features, beauty, plumage, and songs, with the hope that your daily life is made richer.

Chapter 1

Watching Water

Almost certainly you have used a brush to paint at least once, if not many times in your life.

Figure 1.1

You held the brush in hand, and in order to make the hairs cling together and come to a point, you wet it. Without a doubt, if I take a dry brush, dip it in water, and take it out, the hairs will cling together because they are wet, as we are in the habit of saying. But if I simply hold the brush in the water, it is evident that the hairs do not cling at all:

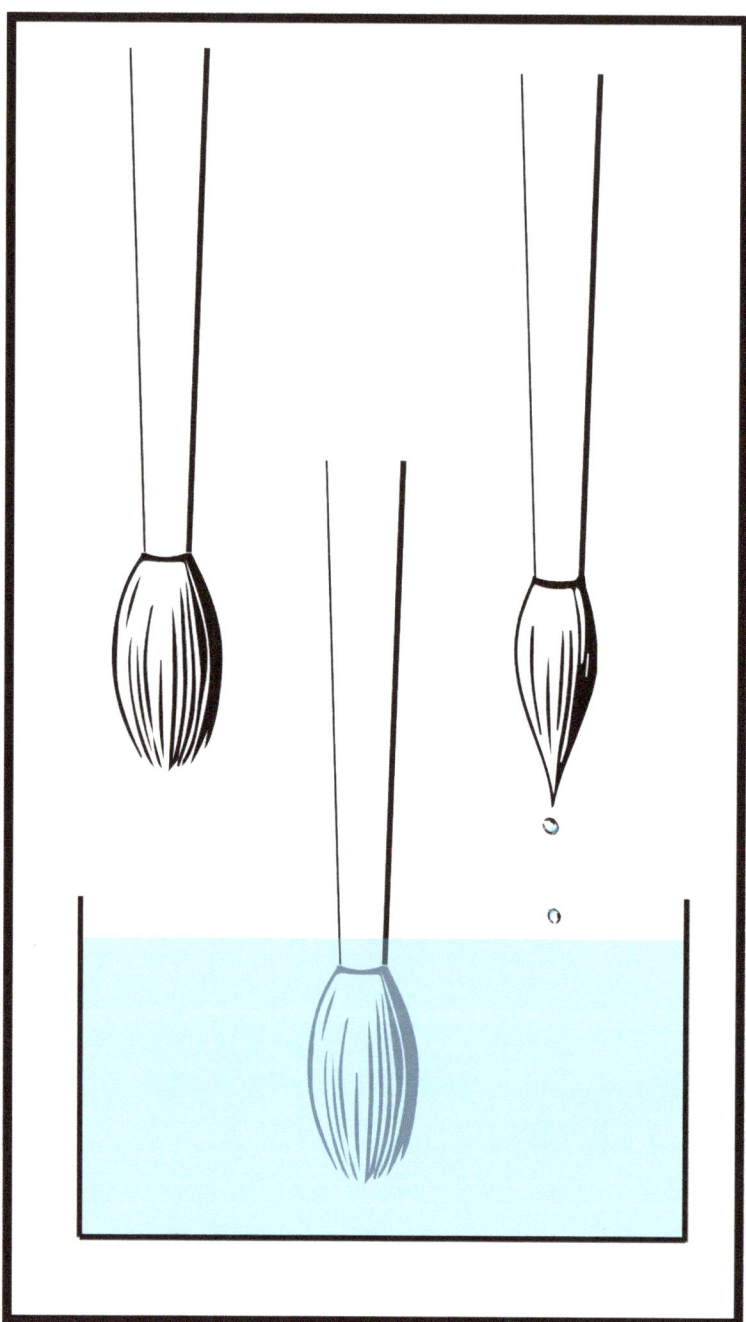

Figure 1.2

So it is not exactly the wetness of the brush that makes the hairs cling together. Rather, it is a phenomenon called *surface tension.*

What is surface tension? All of us had the experience of filling a glass of water which crests above the rim of the glass, but doesn't spill over (Figure 1.3).

How can this happen? Think of a liquid as a collection of molecules which exert attraction forces on each other. Consider Figure 1.4.

Deep inside the liquid a molecule feels equal forces from all directions, but near the surface of the liquid a molecule feels more of a force from inside the liquid than it does from the small number of molecules between it and the surface. Hence, those molecules

Figure 1.3

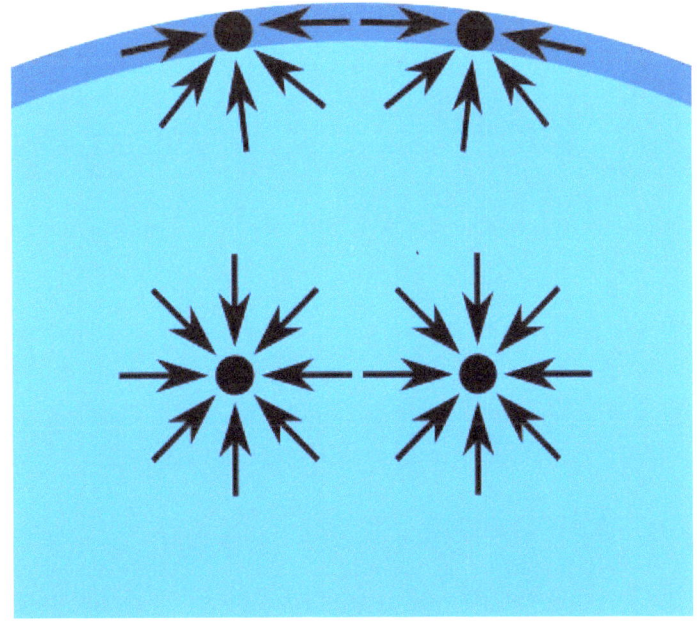

Figure 1.4

near the surface are drawn into the liquid and the surface of the liquid displays a 'curvature'. This property of pulling the surface of a liquid taut is surface tension.

It is this property which allows water striders and other small animals to move on the surface of water.

Figure 1.5: CC BY-SA 3.0, https://commons.wikimedia.org/w/index.php?curid=90142.

Figure 1.6: Basilisk "Jesus Christ" Lizard. By Minetti A, Ivanenko Y, Cappellini G, Dominici N, Lacquaniti F [CC BY 2.5 (http://creativecommons.org/licenses/by/2.5)], via Wikimedia Commons.

As for the brush, we see that when it is immersed in water, the forces balance out, but when it is removed from the water, there is a net force of attraction at the wet part of the brush, resulting in the hairs clinging together.

Isoperimetric Theorem

We can also think of surface tension in terms of energy. Energy is a concept that we all understand intuitively. For instance, a taut rubber band or a wave about to crash are all instances where large amounts of energy are stored, waiting to be released.

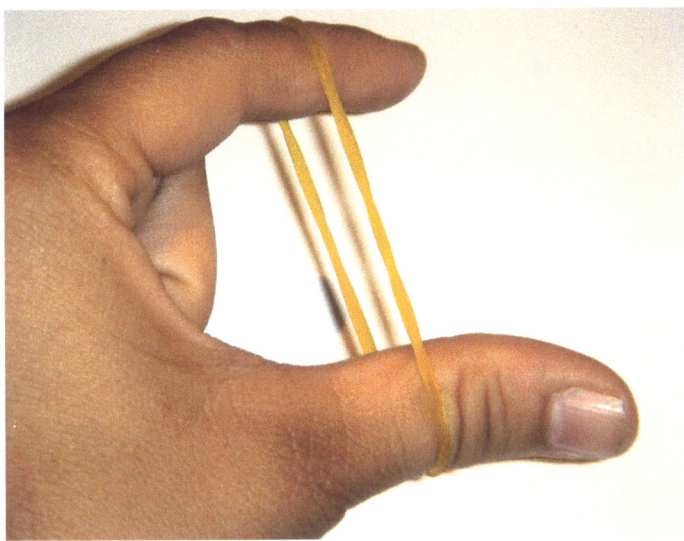

Figure 1.7

Figure 1.8: Large amounts of energy are stored in a wave, and then released as it crashes into the rocks.

A molecule in contact with a neighbor is in a lower state of energy than if it were not in contact with a neighbor - that is, "it wants" to be in contact with its neighbour. The interior molecules have as many neighbors as they can possibly have, but the boundary molecules are missing neighbors. For the liquid to minimize its energy, the number of molecules on the boundary should be minimized. The result is a minimal surface area. Thus, in order to minimize surface tension, *liquids tend to minimize their surface area.*

This immediately leads to mathematics. It is a theorem, called the isoperimetric theorem, that in three dimensions the region having the smallest surface area while enclosing a given volume is a sphere. Together with our knowledge of surface tension, this explains why water droplets assume a spherical shape (Figure 1.9).

Figure 1.9

Namely, the energy due to surface tension is proportional to the surface area, and by the theorem, surface area is minimized for the sphere.

The situation is actually a little more complicated because gravity also has an effect. Thus, it is the size of the body of water which ultimately determines its size; if it is large, gravity dominates and the water flattens; if it is small, surface tension dominates and the water takes on a spherical shape.

We can also explain the odd behavior of the metal mercury. Mercury has a much higher surface tension than water, so even large amounts of it form a sphere:

Figure 1.10: Mercury By Unkky, CC BY-SA 3.0, https://commons.wikimedia.org/w/index.php?curid=1082369.

Figure 1.11: Water

The isoperimetric theorem mentioned earlier, whose statement is that the sphere minimizes surface area subject to a given volume, offers insight into many other phenomena in nature. For example, it tells us why a cat curls up on a cold winter night – to minimize its exposed surface area.

Figure 1.12

Finally, the effect of surface tension are quite pronounced and visible in soap film. As discussed earlier, surface tension leads a soap film to minimize surface area. Such shapes, called *minimal surfaces*, are well-studied mathematical objects with many nice properties. Examples include the Helicoid and Catenoid:

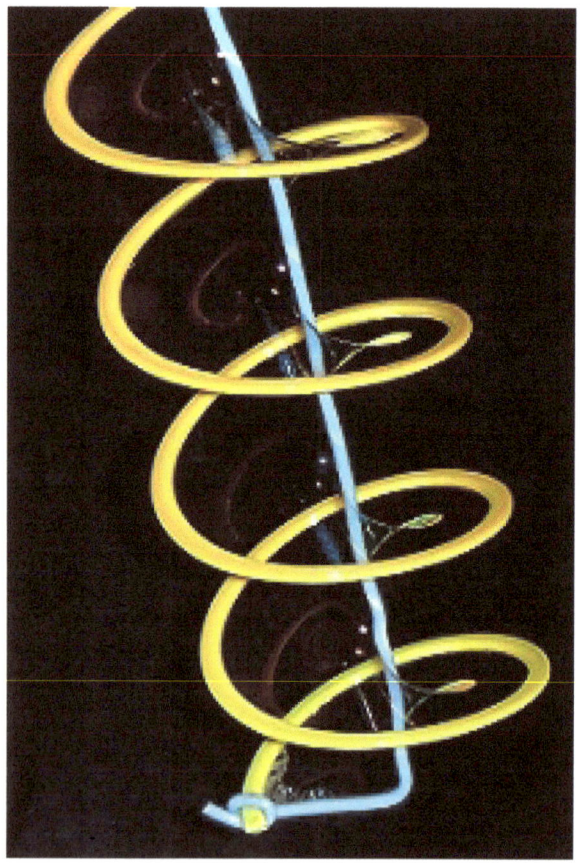

Helicoid.   "Bulle de savon hélicoïde" by Blinking Spirit - Own work. Licensed under CC BY-SA 3.0 via Wikimedia Commons - https://commons.wikimedia.org/wiki/File:Bulle_de_savon_h%C3%A9lico%C3%AFde.PNG#mediaFile:Bulle_de_savon_h%C3%A9lico%C3%AFde.PNG

Figure 1.13

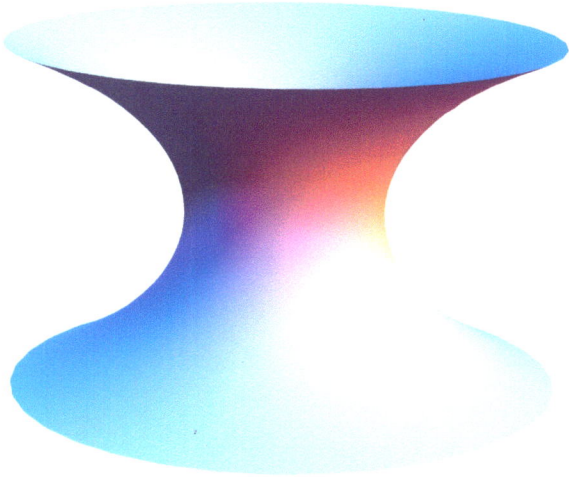

Figure 1.14: Catenoid

Chapter 2

Math in the Garden

"Study how water flows in a valley stream, smoothly and freely between the rocks. Also learn from holy books and wise people. Everything - even mountains, rivers, plants and trees - should be your teacher."

Morihei Ueshiba, founder of the Japanese martial art of aikido

This chapter discusses the amazing fact that the pattern of growth of leaves on the stem of a plant can be explained by a beautiful mathematical theorem stating that the golden ratio is the most irrational number. *Phyllotaxis* is the arrangement of leaves on a stem of a plant. As a branch grows upward, it generates leaves at regular angular intervals that branch out from the stem.

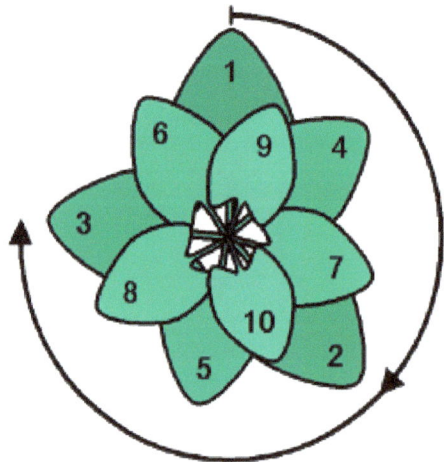

Figure 2.1

If these angular intervals were rational multiples of 360°, then leaves would grow directly above one another, preventing those below them from receiving sunlight and moisture. We illustrate this in the following example, where the angle is 120°.

Imagine that we are constructing the plant. We begin with the stem of the plant, and we place the first leaf:

Figure 2.2

We place the next leaf at an angle of 120° with respect to the first:

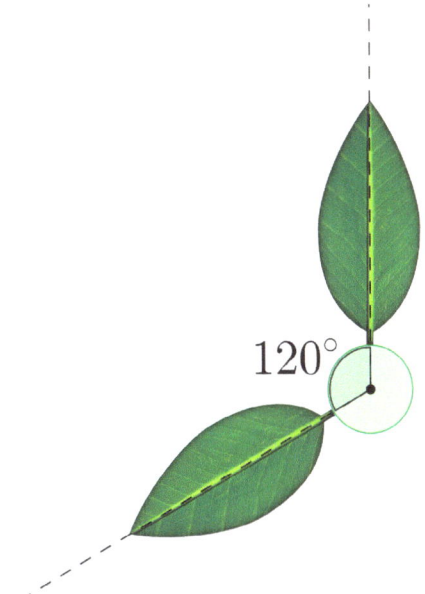

Figure 2.3

And then the next one:

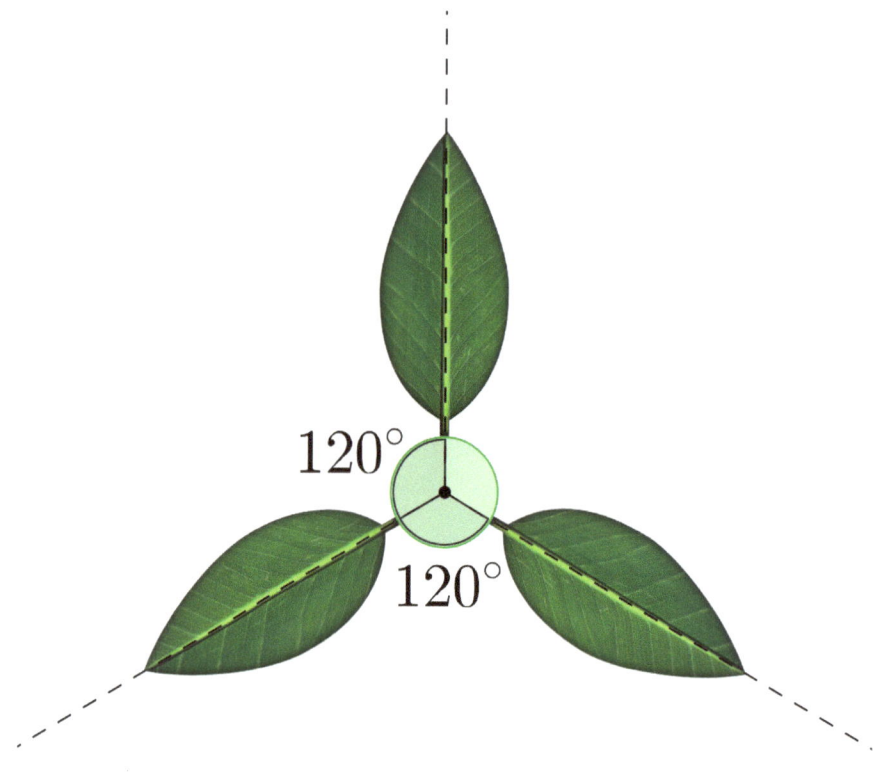

Figure 2.4

Looking at this last figure, we see that a subsequent leaf will grow immediately above the first leaf. As a result, the leaf above would block the sun from reaching the leaf below.

A similar issue arises for any other rational angle; there is bound to be overlap between the leaves[1]. Consequently, a plant should have an angle which is irrational, and maybe even the "most irrational" in the sense that the angle will lead to the least overlap. This angle ought to be as far away from rational as possible. Such an angle does exist(!) and turns out to be called the "golden angle", the angle corresponding to the famous golden ratio. It is the angle for which the ratio of the length of the major arc to the minor arc equals the golden ratio.

---

[1]If the angle is $\frac{p}{q}(2\pi)$ with p, q relatively prime, then the condition that the $(n+1)$st leaf overlaps with the first one is that n is an integer multiple of q. In the case illustrated, the angle $\frac{1}{3}(2\pi)$, so each 4th leaf overlaps with the first.

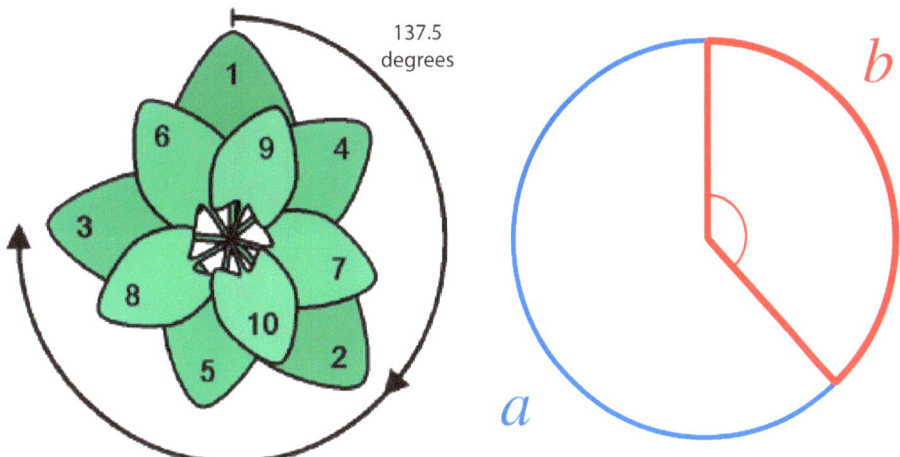

Figure 2.5:   The ratio of the length of the major arc to the length of the minor arc is equal to the golden ratio: $\frac{a}{b} = \phi$. This happens when the angle is about $137.5°$. First image, by The original uploader was Wolfgangbeyer at German Wikipedia - Transferred from de.wikipedia to Commons., CC BY-SA 3.0, https://commons.wikimedia.org/w/index.php?curid=2157671.

Figure 2.5 shows the angle between subsequent leaves of many plants. This is because the golden ratio is the "most irrational number" (more precisely discussed in the appendix). To recap, plants often grow so that the angle formed between consecutive leaves is the golden angle, and this is because the golden ratio is the most irrational number.

This is manifested visually in a wide array of plants and flowers. For many flowers, the number of petals follows the Fibonacci sequence, a sequence of numbers intimately related to the golden ratio. The first few numbers of the Fibonacci sequence are

$$1, 1, 2, 3, 5, 8, 13, 21, 34, 55, \ldots$$

See if you can guess the pattern[2]. The relation between the Fibonacci numbers and the golden ratio is that the ratio of two consecutive Fibonacci numbers approaches the golden ratio:

$$\frac{1}{1} = 1, \frac{2}{1} = 2, \frac{3}{2} = 1.5, \frac{5}{3} = 1.\overline{6}, \frac{8}{5} = 1.6, \ldots, \frac{55}{34} = 1.61765.$$

The golden ratio is

$$\phi = 1.61803\ldots$$

---

[2]Each number is the sum of the previous two.

trillium ~ 3 petals

wild rose ~ 5 petals

Figure 2.6

bloodroot ~ 8 leaves

black-eyed susan ~ 13 petals

Figure 2.7

Chicory ~ 21 petals

gaillardia ~ 34 petals

Figure 2.8

daisy ~ 55 petals   **sunflower ~ 89 petals**

Figure 2.9

The Fibonacci numbers were first conceived by Italian mathematician Leonardo of Pisa, known as Fibonacci, in an attempt to estimate the population of rabbits.

Figure 2.10: https://flic.kr/p/b1fHf, CC2.0

* Appendix

The following gives the precise mathematical statement that the golden ratio is the most irrational number. Readers without the necessary background are urged to skip this section.

Consider a number $\alpha \in \mathbb{R}$. We would like to decide whether a rational number $\frac{p}{q}$ gives a good approximation of α. Of course, we can approximate α as well as we would like in the sense that $|\alpha - \frac{p}{q}|$ can be made arbitrarily small. However, we would like the approximating number to have a small denominator q. At first sight, it seems reasonable to say that $\frac{p}{q}$ gives a good approximation of α if $q \cdot |\alpha - \frac{p}{q}|$ is small. Unfortunately, this does not give a good measure of approximability because every number α can be approximated arbitrary well in this measure: $\forall \alpha, \epsilon > 0$, there exist infinitely many fractions $\frac{p}{q}$ such that

$$q \cdot \left|\alpha - \frac{p}{q}\right| < \epsilon.$$

It turns out that the proper measure of approximability is quadratic approximability. We will say that $\frac{p}{q}$ is a *good* approximation of α if the product $q^2 \cdot |\alpha - \frac{p}{q}|$ is small.

Theorem 2.1 *(A. Hurwitz, E. Borel)*

(a) For any α, there exist infinitely many fractions $\frac{p}{q}$ such that

$$q^2 \cdot \left|\alpha - \frac{p}{q}\right| < \frac{1}{\sqrt{5}}$$

(b) There exists an irrational number α such that for any $\lambda > \frac{1}{\sqrt{5}}$ there are only finitely many fractions $\frac{p}{q}$ such that

$$q^2 \cdot \left|\alpha - \frac{p}{q}\right| < \frac{1}{\lambda}$$

In other words, there exists some irrational number which is badly approximable: this number, the "most irrational", is the golden ratio[3] $\frac{1+\sqrt{5}}{2}$.

Notes

Proofs for the content in the appendix may be found in [FT07].

---

[3]The golden ratio is the unique such number up to the following relation: two numbers α and β are *related* if $\alpha = \frac{a\beta + b}{c\beta + d}$, where $ad - bc = \pm 1$.

Chapter 3

Fractals

Imagination tires before Nature.

Blaise Pascal

Fractals in Nature

The geometry we learn in school is merely the first step towards understanding the geometry of nature. One reason lies in its inability to describe the shape of a cloud, a mountain, a coastline, or a tree. Clouds are not spheres, mountains are not cones, coastlines are not circles, and bark is not smooth, nor does lightning travel in a straight line.

Figure 3.1: Waterfalls are fractal - they exhibit self-similarity at different scales.

More generally, many patterns in nature are so irregular and fragmented that Euclidean geometry of the type learned in school does not provide an appropriate description. The number of distinct scales of length of natural patterns is for all practical purposes infinite.

Figure 3.2: Lightning displays fractal patterns - a part appears like a rescaled version of the whole. Right: by Bert Hickman (en:Image:Square1.jpg) [Attribution], via Wikimedia Commons.

The concept of a *fractal* was introduced by IBM researcher Benoit B. Mandelbrot only a couple decades ago. Expressed in its simplest form, fractals are images in the real world which tend to consist of many complex patterns that recur at various scales.

Figure 3.3: Trees display fractal patterns.

For instance, in the Ramanesco broccoli (Figure 3.4), each little mound is a scaled down version of the whole.

Mandelbrot proposed the idea of a fractal (short for "fractional dimension") as a way to cope with problems of scale in the real world. He defined a fractal to be any curve or surface that is self-similar, meaning that any portion of the curve, if blown up in scale, would appear identical to the whole curve.

Figure 3.4: Romanesco broccoli. https://flic.kr/p/bmdTD4, CC2.0

A huge variety of natural objects are fractal. As shown earlier, waterfalls, trees, brocolli and lightning are. Some more examples are fern leaves, nautilus shells and river deltas.

Figure 3.5: Fern leaves and river deltas are fractal.

Figure 3.6: Nautilus and its shell. The shell is fractal, exhibiting self-similarity.

Additionally, more and more real networks are being discovered to be self-similar fractals. These include the WWW, metabolic network and protein interaction network of H. sapiens.

Fractals in the Body

Many parts of the body have fractal structures. For instance, our lungs (Figure 3.7) and brains (Figure 3.8) are fractal structures.

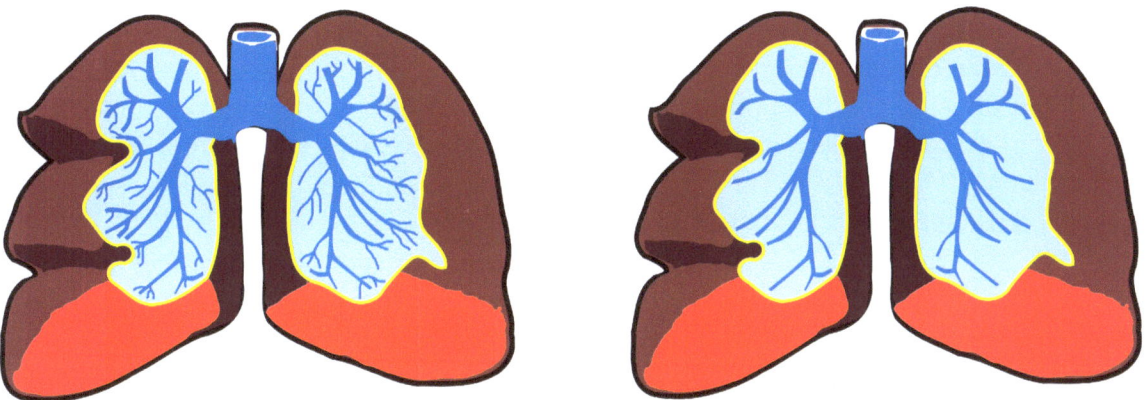

Figure 3.7: Normal (left) and abnormal (right) lungs.

Understanding of their fractal structure (e.g., the fractal dimension to be discussed later) can be used to measure the differences between normal structures and those altered by disease. Our heartbeats seem regular and rhythmical, but when the timing is examined carefully, it turns out to be very slightly fractal. This is very important – our heartbeats are not regular, and this fine variation reduces the wear and tear on the heart drastically. Heart disease can be detected by extreme and arrhythmic fractal behavior. Our lungs, due to their fractal structure, cram the area of a tennis court into the space of just a few tennis balls. Our brain, too, has an obvious fractal structure – it is very wrinkly and convoluted, folding back and forth on itself.

As mentioned earlier, the fractal dimension, to be explained in a later section, can be used to measure the differences between normal structures and those altered by disease. For example, normal lungs have a higher fractal dimension ($D = 1.65$) than hypoxic ($D = 1.53$) or hyperoxic ($D = 1.43$) lungs. As a consequence, the fractal dimension can be very informative.

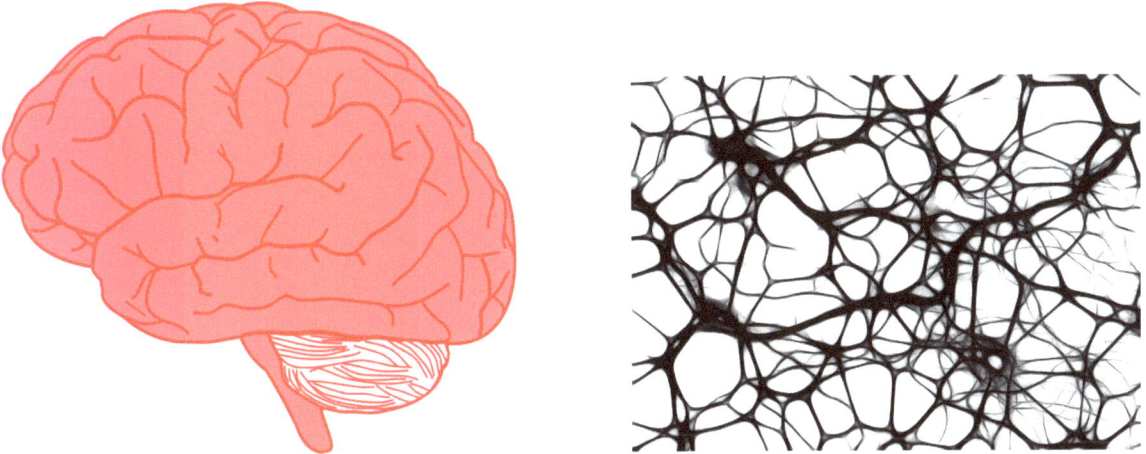

Figure 3.8: Neurons are the cells of the brain. The fractal branching pattern of the neuron's axons and dendrites allows them to communicate with a huge number of other cells. If neurons were shaped like cubes and neatly packed into the brain, one neuron could only connect with at most 6 other cells.

Nature, in general, follows mathematical rules that involve some roughness and a lot of irregularity. For example, complex protein surfaces fold up and wrinkle around towards three-dimensional space in a dimension that is around 2.4 (see section on fractional dimension). Antibodies bind to a virus through their compatibility with the specific fractal dimension of the surface of the cell with which they intend to react. Consequently, many of the current developments and findings in fractal geometry are in work with surfaces.

Understanding of fractals is also used in detecting cancer. The surface structures of cancer cells are wrinkly and display fractal properties which vary markedly during the different stages of the cancer cell's growth. Fractal geometry is being employed in the initial detection of the presence of cancer cells in the body. With the help of computers, mathematical pictures and data are measured and analyzed, revealing whether or not cells are going cancerous. This is done by determining how fractal the cells in the image are. If the cells are too fractal, it spells trouble – There is something wrong with those cells.

As the above discussion suggests, understanding fractals allows for a deep understanding of many natural processes around us.

Fractals in Mathematics

A great variety of fractals show up in mathematics, either by discovery or by construction. One famous example is the Mandelbrot Set. Named for its discoverer, the Mandelbrot Set describes a fantastical shape that displays amazing self-similarity no matter what scale it is looked at and can be rendered with a simple equation.

Here's what I mean. This is the Mandelbrot set:

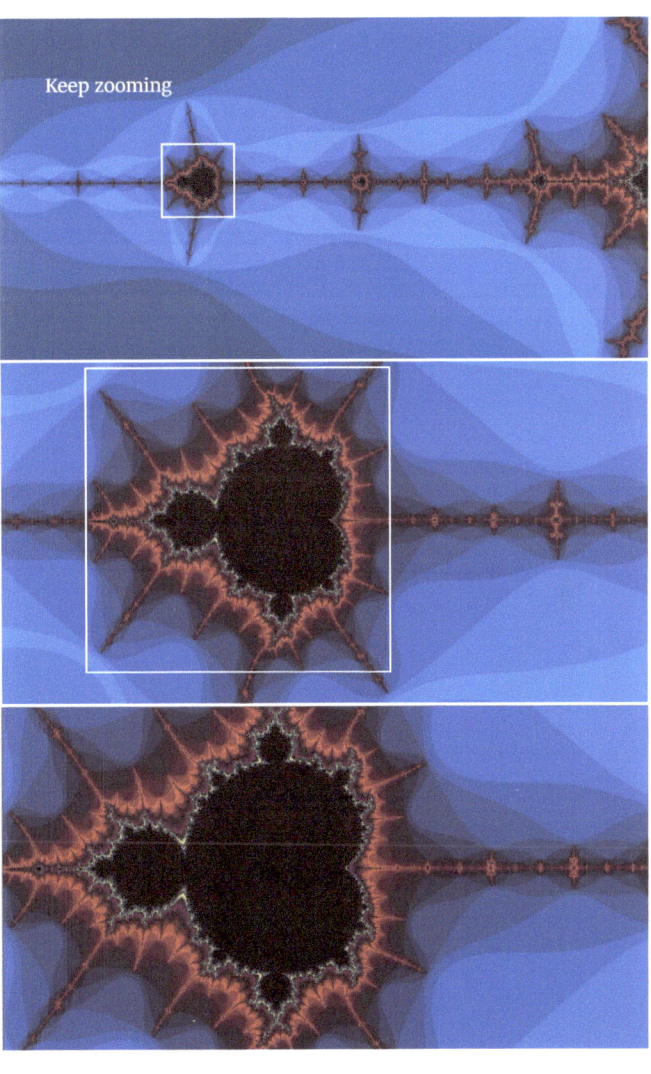

Figure 3.9

Notice the similarity between where we start and where we end.

For exploration of mathematical fractals, we recommend the software XaoS, which allows a user to interactively zoom and play around with the fractal.

A few more mathematical fractals are:

Figure 3.10: The Koch snowflake. "Koch Snowflake 7th iteration" by Wrtlprnft - Own work. Licensed under CC BY-SA 3.0 via Wikimedia Commons.

Figure 3.12: Fractal in three dimensions.

Figure 3.11: The sierpinski triangle.

Fractional Dimension

In this section, we shall discuss the bizarre notion of a fractional dimension. Interest in this topic, as in many exotic mathematical notions, arose from practical considerations. As alluded to earlier, in the medical sciences, an abnormal fractal dimension is symptomatic of disease, and consequently understanding and being able to measure fractional dimension is critical in diagnosis.

To illustrate the idea, imagine that a geographer would like to measure the length of the coast of Britain.

With a ruler of length 200km the geographer measures something like this:

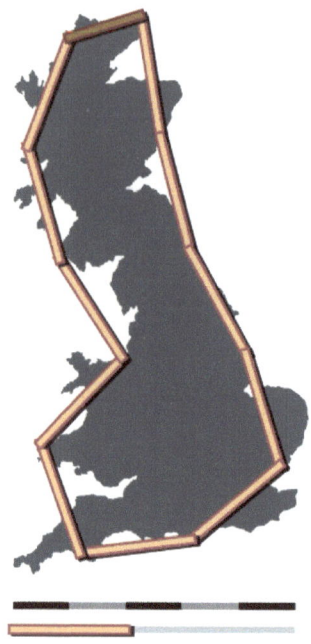

which is about 12 ruler-lengths. This gives an estimate of $12 \times 200 = 2400$km for the coast of Britain.

Next, using a ruler of length 100km, half the size, the geographer obtains a measurement like this:

which is about 28 ruler-lengths. The estimate now is $28 \times 100 = 2800$km. Using an even smaller ruler of length 50km,

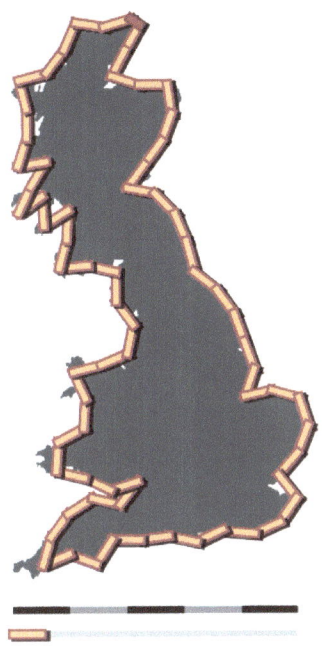

measuring 69 ruler-lengths, or $69 \times 50 = 3450$km. Notice that as the length of the measuring stick is scaled smaller and smaller, the total length of the coastline measured increases radically. Consequently, if another, fellow geographer, also measured the length of the coast of Britain, but used a different scale, he would completely disagree with the first geographer. This is a feature of the fractality of the coast of Britain.

To explain this in concrete terms, the coast of Britain as seen from Earth's orbit looks generally jagged and unsmooth, yet you can still pick out short stretches that appear from that distance to be simple curves or lines. If you zoom in to an altitude of 5000 meters, those previously small and apparently smooth stretches of coast reveal themselves to be intricate and jagged, with inlets and rocky outcrops; although there will, again, be visible some seemingly smooth short segments. Successive magnifications reproduce the same phenomenon of smooth sections turning out to be rough as we zoom in. A photo taken ten centimeters from the edge of a clod of dirt has the same relative degree and type of jaggedness and roughness – of complexity – as a photo of the entire coastline of Britain taken from space. This is what is meant by fractality.

In the terms of the measurement system specified above, the *fractal dimension* of a coastline quantifies how the number of rulers required to measure the coastline changes with the scale of the measuring stick. For example, if we halve the length of the measuring stick, does the number of measuring sticks needed to measure the region double? Triple? Quadruple? Maybe something else...?

The answer is the dimension of the object[1]. For an illustration, consider some

---

[1]Strictly speaking, the dimension is this ratio as the ruler length approaches zero. This statement

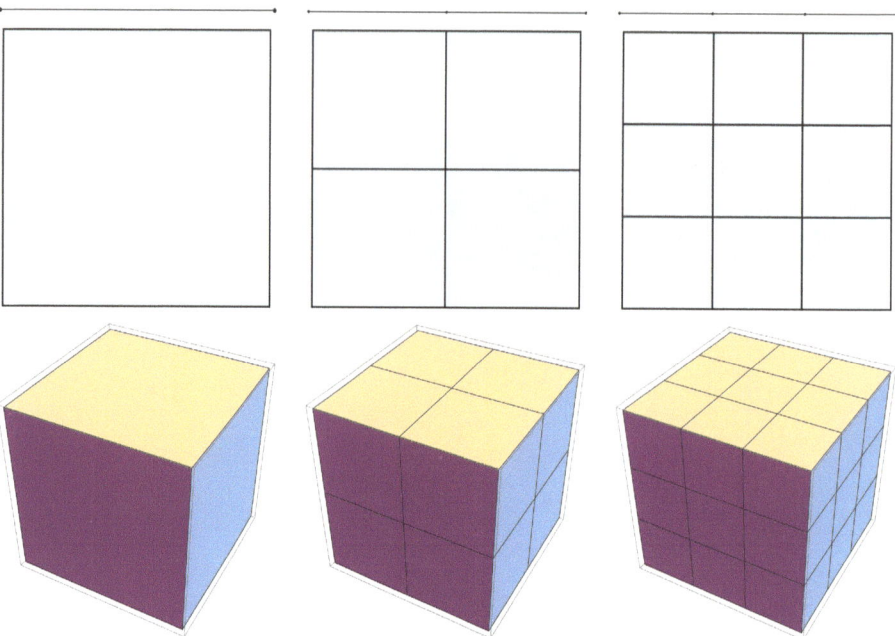

familiar objects such as line segments, squares and cubes.

Measuring a line segment with rulers is completely straightforward: if we use rulers of length 1 it takes only 1 ruler. Measuring it with rulers of length 1/2 takes 2 rulers. Measuring it with rulers of length 1/3 takes 3 rulers. More generally, measuring with rulers of length $1/n$ requires n rulers.

Next consider the square. To measure its area with squares of side-length 1 takes just 1 square. To measure its area with squares of side-length 1/2 takes 4 squares – not 2. To measure its area with squares of side-length 1/3 requires 9 squares. We see then that the number of squares needed to measure the area of the square grows as "the square" n^2.

Finally, consider the cube. To measure its volume with cubes of side-length 1 takes just 1 cube. To measure its volume with little cubes of side-length 1/2 takes 8 cubes – not 2 or 4. To measure its volume with little cubes of side-length 1/3 requires 27 cubes. Evidently, the number of cubes needed to measure the cube grows as "the cube" n^3.

The exponent d in n^d is what we intuitively consider to be the dimension. The line segment is 1-dimensional, so we have growth of rate n. The square is 2-dimensional, we have growth of rate n^2. And the cube is 3-dimensional, implying that it has growth of rate n^3.

We can summarize this by saying that

The dimension tells us how many new pieces we see when we look at finer resolution.

For fractals, the dimension d is not necessarily an integer. As we see in the example of the coast of Britain, the number of measuring sticks needed does not double or

___

can be made precise using the tools of real analysis.

quadruple:

$$12 \to 28 \to 69.$$

Rather, the growth rate is always something in between. Hence we say that the dimension of the coast of Britain is between 1 and 2, and is a fractional dimension.

For the koch snowflake (Figure 3.10), we can calculate the dimension precisely – it is the non-integer $\frac{\log(4)}{\log(3)}$.

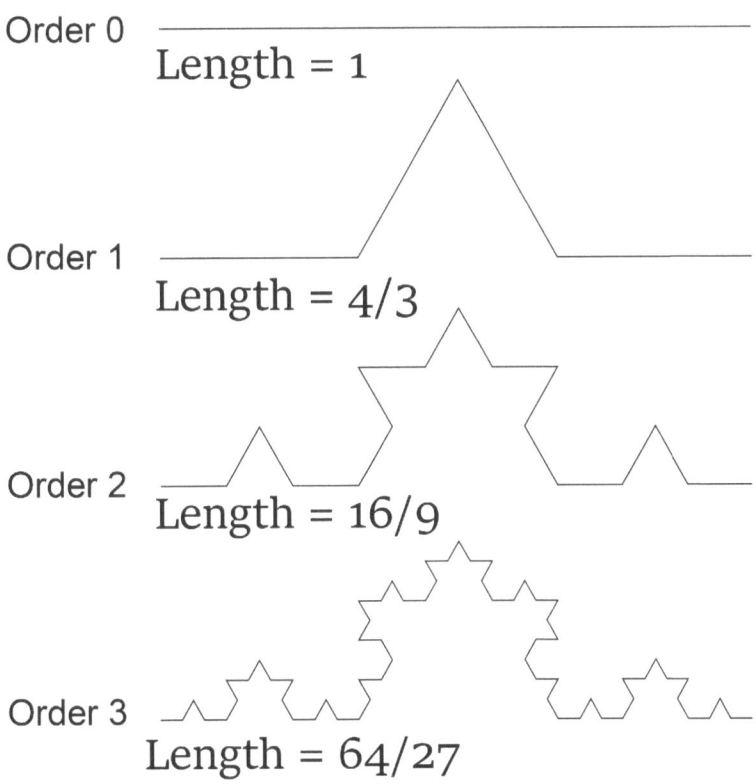

Figure 3.13:   The fractal dimension of the koch snowflake is $\frac{\log(4)}{\log(3)}$.

Consequences of Fractality

In previous sections, we saw many examples of fractals and discussed their main features: self-similarity at different scales, coarseness and fractional dimension. Let us now discuss some implications of these facts.

1. **There is no unique "correct" value for relevant measurements**: the value of the measurement may depend on the resolution, just as we saw in measuring the coast of Great Britain. In particular, different people can arrive at different results in their measurements because they are using different resolutions ("ruler sizes").

2. **Measuring a value at one resolution is not useful**: we must understand the scaling relationship of the object in order to make useful statements, which requires measurements at different resolutions.

3. **Many statistical measures do not make sense**: for example, the "average", say the average size of an artery in an arterial tree will continue to decrease as more measurements are made because there is an effectively never-ending supply of smaller and smaller arteries.

Chapter 4

Rainbows

It is much better to understand the rainbow than to chase the rainbow. In this chapter, we will do just that.

Figure 4.1: Primary rainbow

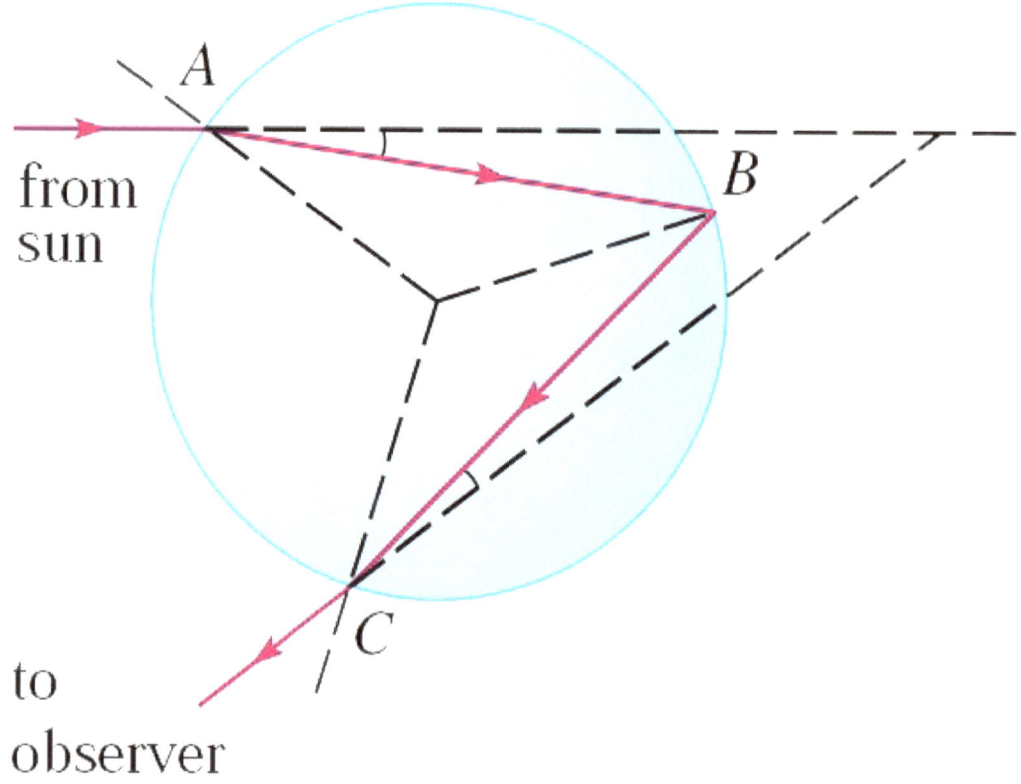

Figure 4.2

Rainbows are created when raindrops scatter sunlight. A ray of light hits a raindrop (A). It bends as it enters the raindrop and continues moving inside of the raindrop until it hits the boundary at B. At this point, part of the ray of light passes through and part reflects inwards. The part that passed through we don't see. The part reflected inwards continues until it reaches another boundary (C). The light that exits the raindrop at this point and reaches our eyes forms the primary rainbow.

This model explains:

1. Why the rainbow has different colors.

2. Why the rainbow appears at the same spot in the sky (why you don't chase the rainbow!)

3. The formation and location of a secondary rainbow.

To clarify #2, take a look at Figure 4.3.

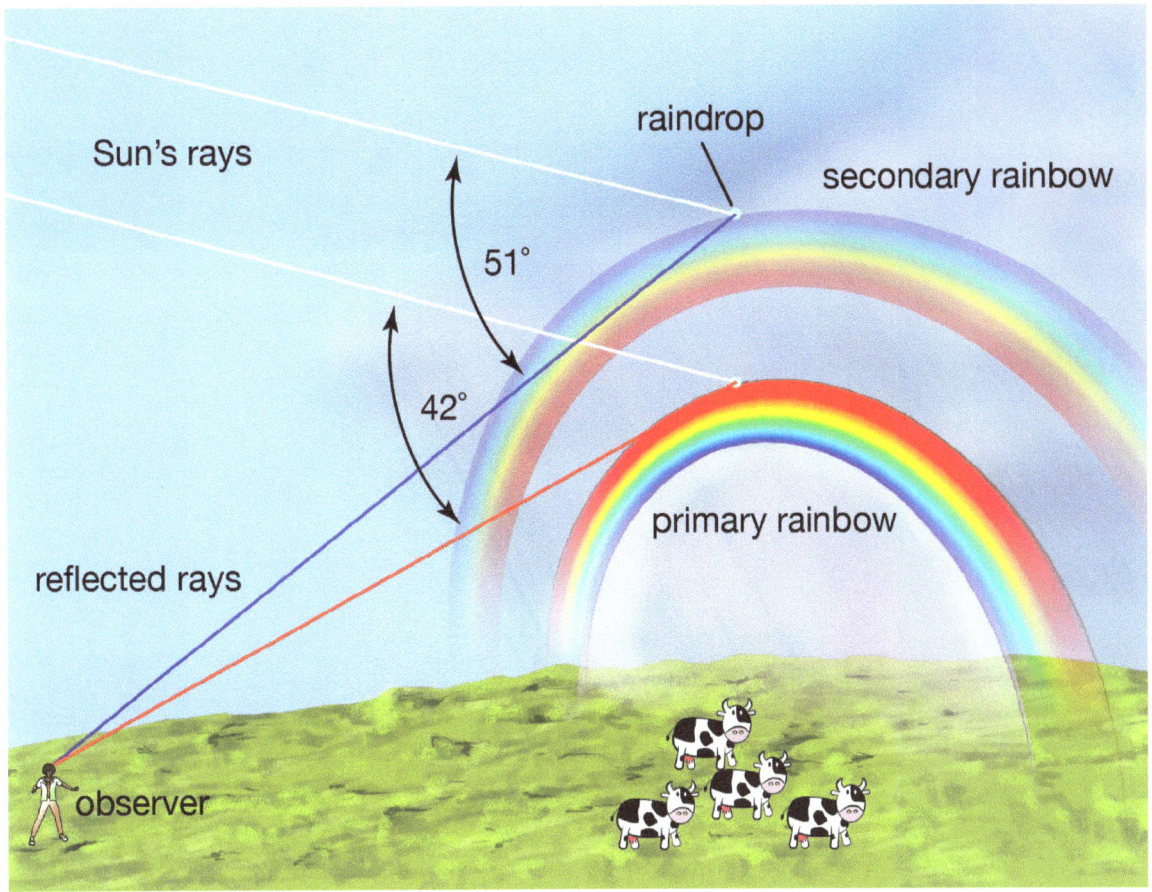

Figure 4.3

We always see the primary rainbow at about 42°. This angle is called the rainbow angle. We see the rainbow at this angle because it occurs when the angle of deviation is minimal, and it is equal to

$$180° - \text{minimum angle of deviation}$$

See Figure 4.4.

How come light takes the angle of minimum deviation, you might ask. Well, around the minimum angle of deviation, the math shows that there is an unusually high concentration of light there – so it's not that light really takes the path of least angle of deviation, but that an overwhelming majority of the light rays do, and that is what our eyes see.

Colors of the Rainbow

The rainbow has different colors because different colors of light bend more strongly than others when passing from air to water. This is illustrated in the famous glass prism experiment (Figure 4.5).

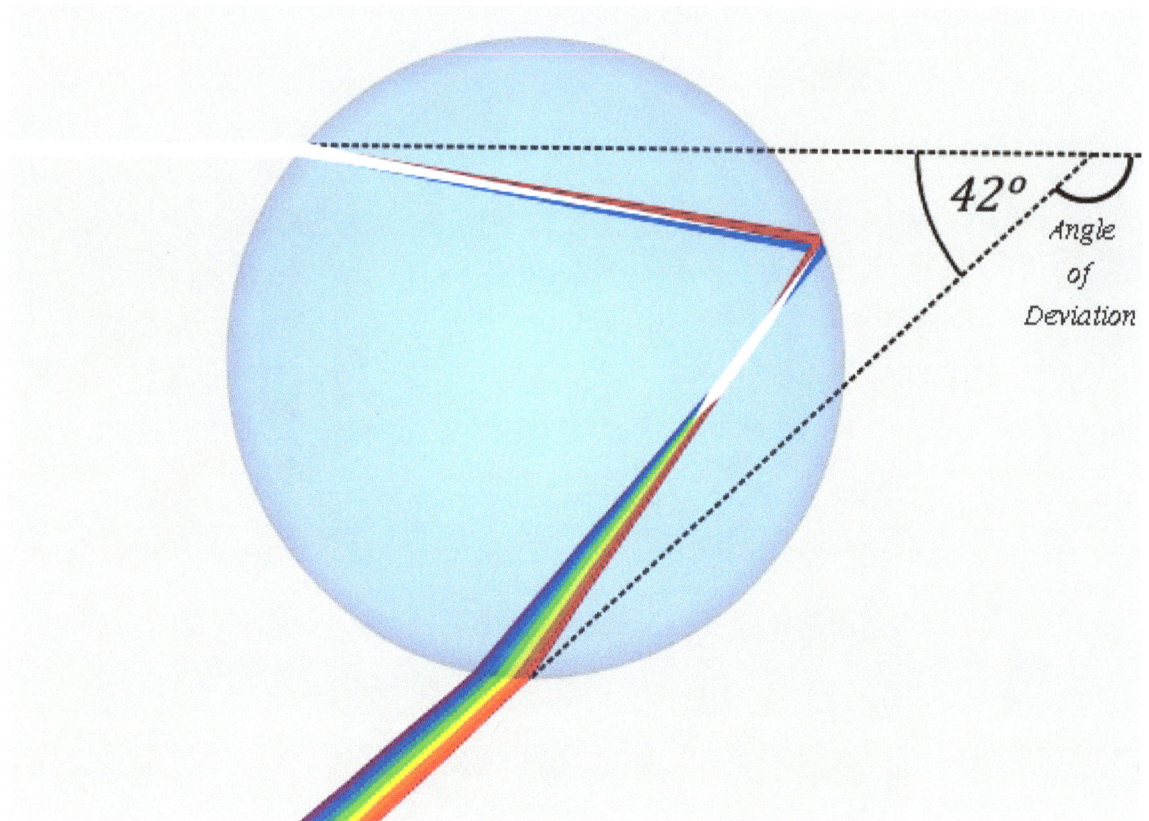

Figure 4.4: Rainbow Angle and Angle of Deviation

Figure 4.5

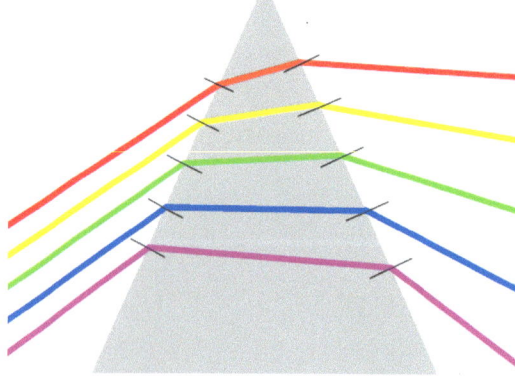

Figure 4.6: Different colors of light refract at different angles.

In our situation, the prism is the raindrop. The minimum angles of deviation for the different colors differ, resulting in different rainbow angles.

By the way, this is also the reason that the sky is blue and during a lunar eclipse, the moon appears red. The reader is urged to provide an explanation based on the differential bending of light as a function of color.

Figure 4.7

Double Rainbow

Next, let us explain the formation of a double rainbow. Anytime there is a rainbow, a second, fainter rainbow appears above it:

Figure 4.8: Double rainbow By Eric Rolph at English Wikipedia - English Wikipedia, CC BY-SA 2.5, https://commons.wikimedia.org/w/index.php?curid=2406447

This rainbow has its colors in reverse order. This is explained as follows. Just as before, light enters a raindrop. This time, however, it reflects twice inside of the raindrop before exiting:

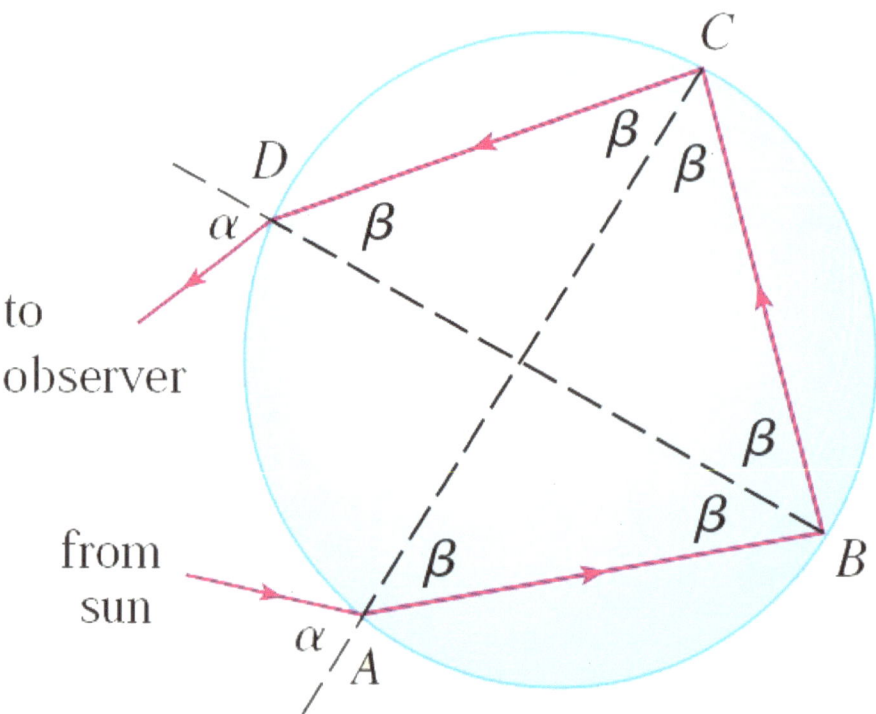

Figure 4.9

One can calculate, again using the principle that the angle of deviation is minimized, that the rainbow angle of the secondary rainbow is about 51°.

This principle can be extended: there are triple, quadruple, etc. rainbows and their angle can be predicted. In nature, triple and quadruple rainbows are extremely rare (though they have recently been documented and photographed). The reason is that the bows become fainter as their order increases and are visually closer to the sun. But in the lab, these are readily made.

Chapter 5

Crystallography

Crystals play a role in many subjects, among them mineralogy, inorganic, organic and physical chemistry, physics, metallurgy, materials science, geology, geophysics, biology and medicine. This pervasiveness is perhaps better understood when it is realized how widespread crystals are: virtually all naturally occurring solids are crystalline. A mountain crag normally is made up of crystals of different kinds, while an iceberg is made up of many small ice crystals. Virtually all solid inorganic chemicals are crystalline, and many solid organic compounds are made up of crystals, among them benzene, naphthalene, polysaccharides, proteins, vitamins, rubber and nylon. Metals and alloys, ceramics and building materials are all made up of crystals. The inorganic part of teeth and bones is crystalline. Hardening of the arteries and arthritis in humans and animals can be traced to crystal formation. Even many viruses are crystalline.

Formally, *a crystal consists of atoms arranged in a pattern that repeats periodically in three dimensions [Bar52].*

See figure 5.1. Substances may be "crystalized", meaning that one forms a crystal by creating a periodically repeated pattern of the molecules of the substance. This allows to determine the precise molecular structure of the substance using X-ray and neutron techniques. This knowledge allows new pharmaceuticals and functional materials to be designed.

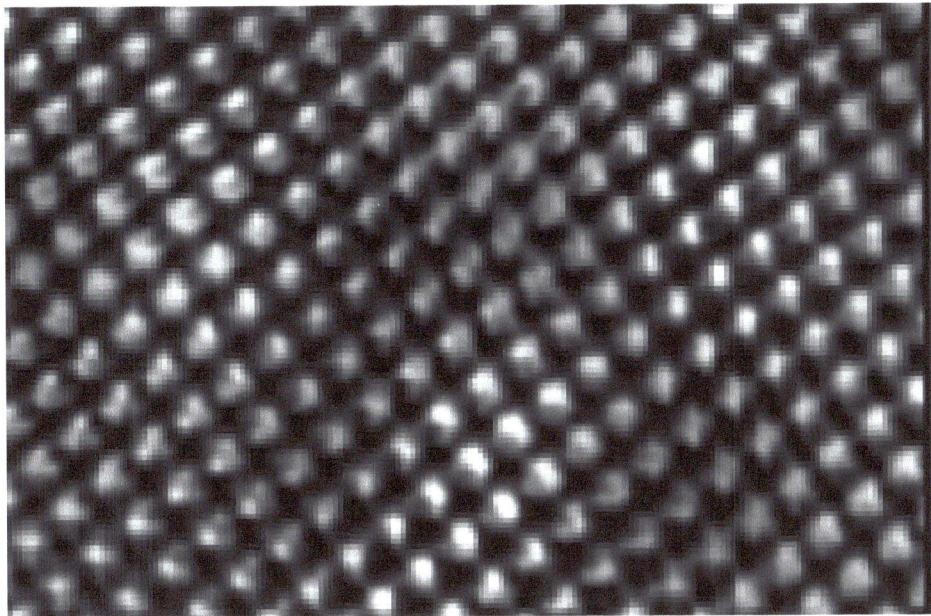

Figure 5.1: The periodic structure of a crystal, shown here under a microscope. Original by Erik Wetterskog et al. [CC BY 3.0 (http://creativecommons.org/licenses/by/3.0)], via Wikimedia Commons.

Figure 5.2: Salt crystals.

Shape of Crystals

Symmetry

Many crystals have not only smooth faces, but also a regular geometric shape. Fig. 5.3 shows a Chromium Alum crystal forming an octahedron. The beautiful geometric shape is a result of the internal regularity of the crystal.

Figure 5.3: Chromium Alum. By Ra'ike (Own work) [GFDL (http://www.gnu.org/copyleft/fdl.html) or CC BY-SA 3.0 (http://creativecommons.org/licenses/by-sa/3.0)], via Wikimedia Commons.

Figure 5.4: Quartz crystal.

To understand this a little better, remember that a crystal is a collection of molecules which is repeated over and over again in space (see for instance the left part of Figure 5.5). The unit which is being repeated is called the *unit cell*.

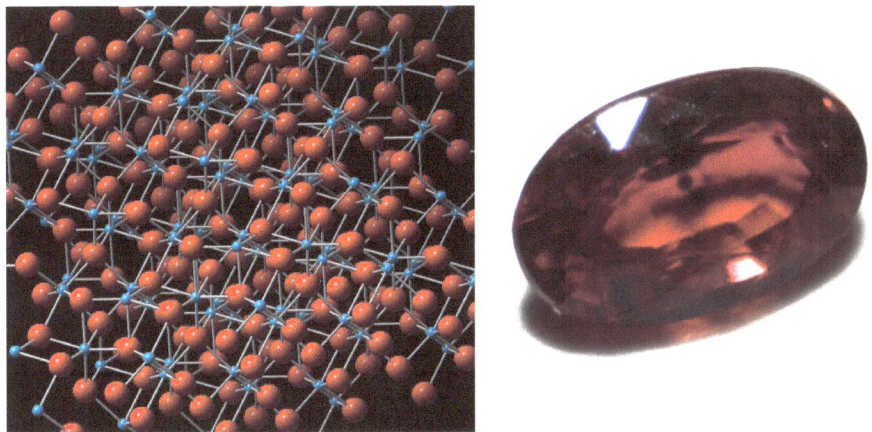

Figure 5.5: The molecular (crystal) structure of ruby. Right image by Humanfeather - Own work, CC BY 3.0, https://commons.wikimedia.org/w/index.php?curid=6969673.

Consider the molecular structure of ordinary salt (Figure 5.6). Several unit cells are shown in Figure 5.6.

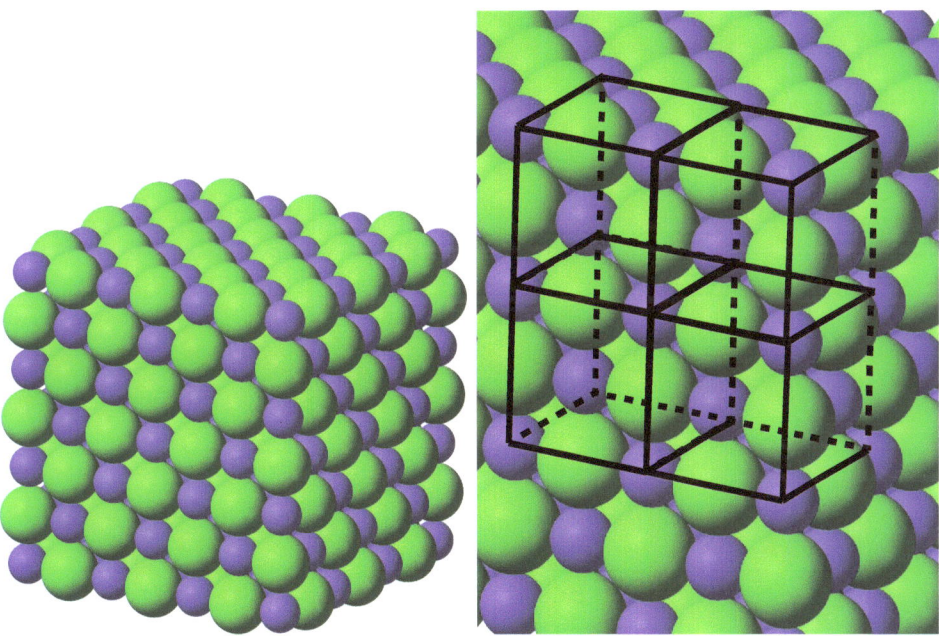

Figure 5.6: Left: Molecular (crystal) structure of sodium chloride (table salt). Right: Some unit cells of sodium chloride. By Benjah-bmm27 (Own work) [Public domain], via Wikimedia Commons.

Imagine replacing each unit cell with a single point at its center. We now obtain a collection of points repeating in space. Such a structure is called a *lattice*. For example, in two-dimensional space, the grid (Figure 5.7) is a lattice.

There are various lines which pass through at least two points of the lattice. These are called *lattice lines*. Similarly, there are planes which pass through at least three of the points of the lattice – these are called *lattice planes*.

The external shape of a crystal, i.e., what we actually see with our eyes, reflects the underlying structure of this lattice. A more precise version of this statement is

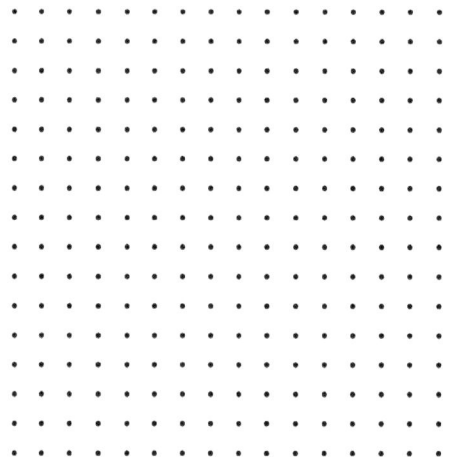

Figure 5.7: The grid, an example of a lattice.

every face of a crystal is parallel to a lattice plane and every edge is parallel to a lattice line.

For example, the lattice for salt is just the three-dimensional grid, and this is consistent with the crystals of salt having a cubic shape whose faces are parallel to some of the lattice planes, and whose edges are parallel to some of the lattice lines.

Of course, the converse of the statement above is not true, as there are many more lattice planes and lines than faces and edges. Interestingly, a given type of crystal can take on different forms (e.g., some crystal might have a cubic version, a prismatic version, etc.) and these would all still have faces and edges coming from the lattice. Moreover, these faces and edges are almost always parallel to the "simplest" lattice planes and lines[1].

Anisotropy and characteristic properties of crystals

The following are a few properties that are characteristic of crystals:

- If some crystals (e.g. NaCl, aka salt) are split, the resulting fragments have similar shapes with smooth faces – in the case of NaCl, small cubes (Figure 5.2). This phenomenon is known as cleavage, and is typical only of crystals.

- Crystals can display directional optical properties. The color of the crystal might depend on the angle at which it is viewed. That is, if I hold the crystal, it might be orange, but as I rotate it in my hand, it will appear purple. This phenomenon is called *pleochroism* (Figure 5.8). In addition, crystals can show a double image, a property called *birefringence* (Figure 5.9).

- Crystals can display directional hardness. For example, when a crystal of kyanite is scratched parallel to its length with a steel needle, a deep indentation will be made in it, while a scratch perpendicular to the crystal length will leave no mark

---

[1]The so-called Miller indices have small integer coordinates

- Crystals can display directional thermal conductivity. If one face of a gypsum crystal is covered with a thin layer of wax and a heated metal tip is then applied to that face, the melting front in the wax layer will be ellipsoidal rather than circular, showing that the thermal conductivity is greater in one direction than in the other.

Figure 5.8: **Pleochroism.** By Mauswiesel (Own work) [Public domain], via Wikimedia Commons.

Figure 5.9: Calcite displaying birefringence.

The phenomena in which the crystal acts as one substance from one angle, but as a different substance from another angle, is called *anisotropy*.

Anisotropy can be explained mathematically as follows. Imagine that a beam of light is shot at the crystal. For simplicity, assume that the atoms are arranged in a 2-dimensional grid, rather than a 3-dimensional one. If we shoot a beam of light in parallel to an axis of the grid, as in Figure 5.10 we will see a completely different result than had we done so in another direction, say diagonally (Figure 5.11).

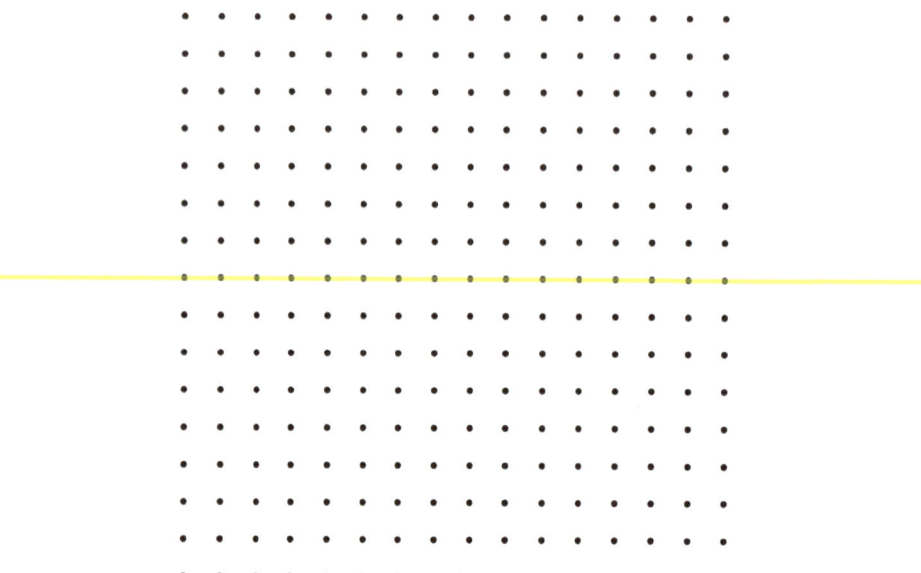

Figure 5.10

Figure 5.11

The directional hardness can be similarly explained. In one direction, the apparent density of atoms is larger than at another.

Limitations of symmetry of crystals

As explained earlier, the internal, microscopic symmetries of crystals manifest themselves in the macroscopic appearance of the crystal, such as the directions of its faces and edges. In particular, symmetries of the lattice give rise to symmetries of the crystal.

Initially it might appear that there could be a limitless variety of symmetry elements. This is, however, not the case.

For the crystal to have an n-fold rotational symmetry is to say that if we rotate it at $(1/n)$-th of a full rotation, the end result should look the same. The unit cells must also have the same symmetry. Hence each unit cell must also have n-fold rotational symmetry. As an example, consider again the cubic unit cells of table salt (Figure 5.6), which have 4-fold symmetry.

Since the unit cells must fill up three-dimensional space, this means in particular that the crystal yields a tiling of two-dimensional space (obtained by restricting our attention to a plane). This means that we must have a tiling of the plane with shapes having n-fold symmetry. This immediately restricts the possible values of n. You have never seen a floor tiled with five sided tiles, because this is not possible (see Figure 5.12).

Figure 5.12: The plane cannot be tiled with pentagons. Unfilled gaps are left.

This also means that a sevenfold rotational symmetry is not possible, etc. The mathematical equation that tells us which tiles are possible is

$$2\cos(2\pi/n) = m, \quad (m \text{ an integer})$$

The only possible solutions are $n = 1, 2, 3, 4, 6$. The only tiles, and therefore the only shapes permissible for the sides, or faces of unit cells that can satisfy the necessary criteria are (1) parallelograms, (2) rectangles, (3) triangles, (4) squares, and (5) hexagons. Any other shapes having more sides on a face are impossible in crystals.

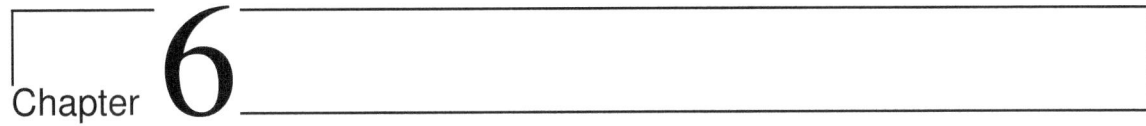

Chapter **6**

Optimization

> "Nothing at all takes place in the universe in which some rule of maximum or minimum does not appear."
>
> Leonhard Euler, great mathematician and physicist

What is mathematical optimization?

To explain what mathematical optimization is, it is easiest to start with a hypothetical example. Imagine an airline company that must plan out its business. First, to make profit, the airline must sell tickets. Pricing the tickets thoughtfully will determine profit: if the tickets are too expensive, consumers will fly with competing airlines; if the tickets are too cheap, the airline will suffer financial losses. Flying airplanes is costly. Flying too many is expensive, but also allows to fly more passengers and therefore sell more tickets. One should consider how to select routes for the planes. Since fuel is expensive, the routes should be efficient. They should also service as many customers as possible. Another concern is that one cannot land many planes at the same time in a given airport, so the schedule of flights is important. This is just the tip of the iceberg. Many other factors must be taken into account, some of which affect one another, and the question of how to manage such an enormously complex problem arises. The answer is mathematical optimization.

The problem is recast into mathematical form with collected data such as fuel cost, distances and consumer demand. Then an efficient algorithm, i.e., a mathematical procedure, is used to solve or approximate the solution, giving answers to all above questions with the goal of maximizing profit. Specialists in the field of optimization look to find accurate, quick and robust (the quality of being insensitive to errors in the collected data) algorithms to such problems. They also seek to understand if such algorithms are possible, how many solutions exist, and to understand the nature of the solutions.

It is important to note that the power of abstraction of mathematics is made full use of here. The specialist in optimization explains how to solve a family of abstract problems. For example, one specifies fuel costs to be some variable f, conditioned to be positive. The exact data is not used to solve the abstract problem, and therefore neither the inner workings of the airline company, nor the data which will be eventually fed in, are needed when working with the abstract problem. In fact, the same family of problems (e.g., a "linear optimization problems") and solution method (e.g., the "simplex algorithm") may be used in a huge number of settings, from menu planning in large restaurants, to optimization of oil refinery outputs.

It is not much of an exaggeration to say that the field of Machine Learning is a subfield of optimization. The basic problems are always to form predictions while minimizing the rate of error. Applications are in abundance – classifying images, spelling-correcting software, spam detection, route planning, trend prediction and more.

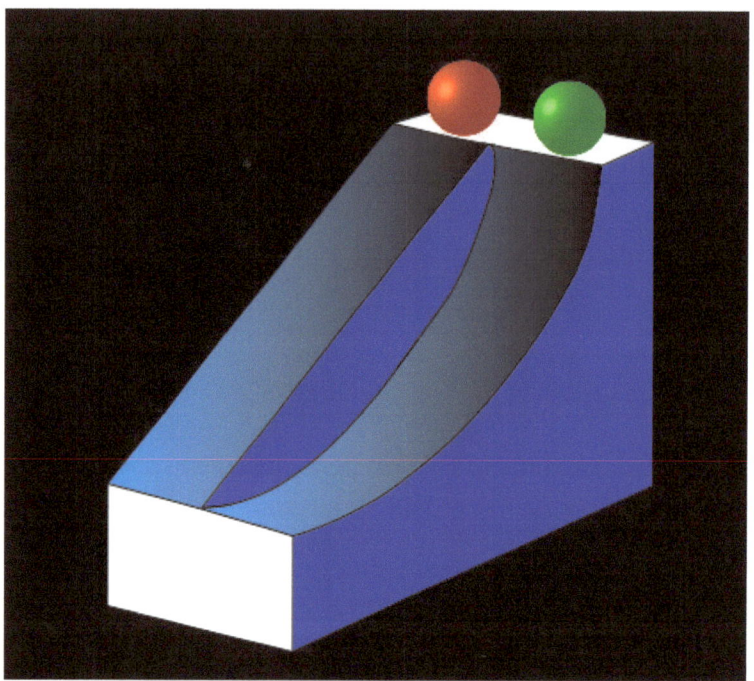

Figure 6.1: The fastest route for a ball to reach the bottom of a hill is not actually the straight line; it is the curve on the right, called the *Brachistochrone*.

Optimization in nature

Nature, too, optimizes. In fact, it is believed that all nature does is optimize. In physics, pretty much all laws of physics, starting from Newton's laws of motion and more modern theories (e.g., Feynman's path integral formulation of Quantum Mechanics) are variations on the *principle of least action*, which states that

The motion taken by a mechanical system is the one for which the quantity called "action" is optimized[1].

One illustration of this principle is that light follows the path of shortest optical length connecting two points, where the optical length depends upon the material of

---

[1]This is the moral of the principle. There are some technical subtleties, mainly that the action is made stationary.

the medium (Figure 6.2). This explains the bending of light as it passes from one medium to another.

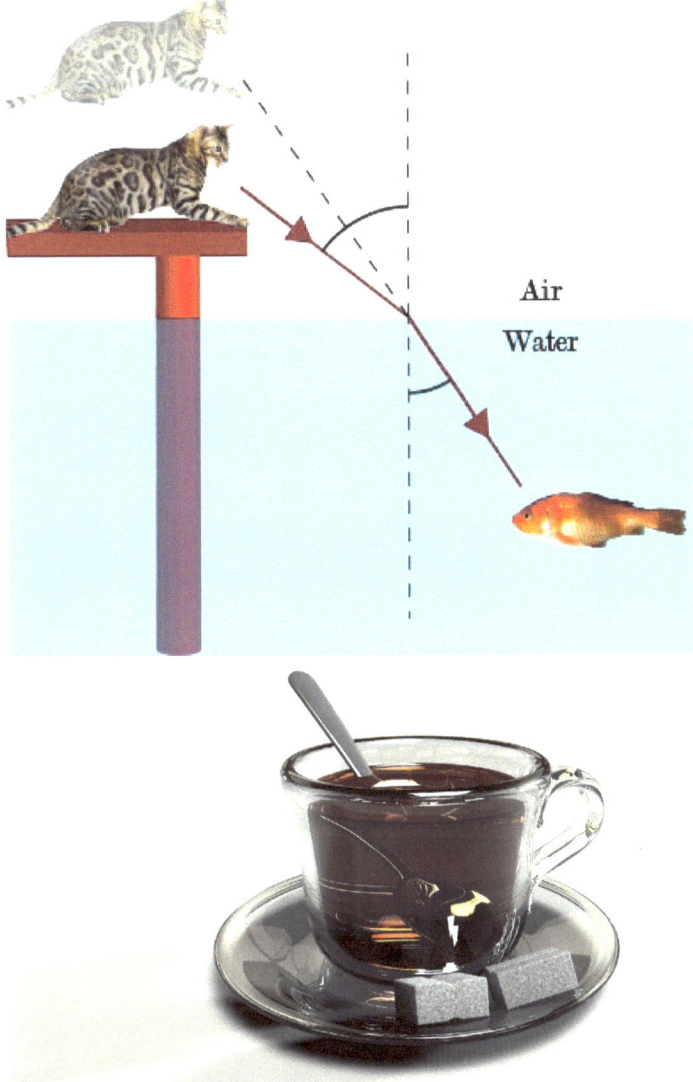

Figure 6.2: Snell's law is a physical law describing the bending of light when passing between two different media. It is directly explained by the principle "Light follows the path of shortest optical length connecting two points".

Another example of optimization in nature was the subject of Chapter 2, in which the growth of the leaves on the stem of a plant is often such that there is a minimal amount of overlap, the goal being to maximize the amount of sunlight absorbed by the leaves.

Mathematical Optimization

In optimization, one is interested in minimizing or maximizing a mathematical function of some sort. It is helpful for intuition to think of the function as a height above sea-level, and the parameters that are being varied in order to optimize the function as geographic location. Thus in the following figure, we would be interested in the locations of the peaks and valleys, and their heights.

Figure 6.3: Plot of a mathematical function.

Local minima are points which are lowest in their vicinity, like valleys, though they might not be the lowest over the whole region being considered. Global minima are locations which are the lowest over the whole region.

One intuitive method for finding minima is called "gradient descent". Here, we imagine a person starting off at some region, and this person takes the path that has the steepest decline. Eventually, the person is expected to reach a valley (a minimum). Unfortunately, they may get stuck at a local minimum instead of the global minimum.

It is a common problem in optimization algorithms that they can guarantee only local minima, rather than global ones. For general hard problems, there is provably no hope for a guaranteed optimal solution. However, it is often the case that specific algorithms, though lacking guarantees and applicable only to specific problems, perform well in practice. Such algorithms are called "heuristic algorithms", as their method is based on a heuristic rather than rigorous and reliable effectiveness.

Nature-inspired optimization algorithms

Ants teach us how to find the right path

Network optimization problems are optimization problems that involve a system with possibly intricate interconnections. For example, consider a phone network. At any given time, a message may take a certain amount of time to traverse each line (due to congestion effects, switching delays, and so on). This time can vary greatly minute by minute and telecommunication companies spend a lot of time and money tracking these delays and communicating these delays throughout the system. Assuming a centralized switcher knows these delays, there remains the problem of routing a call so as to minimize the delays. So, in Figure 6.4, what is the least delay path from LA to Boston? How can we find that path quickly?

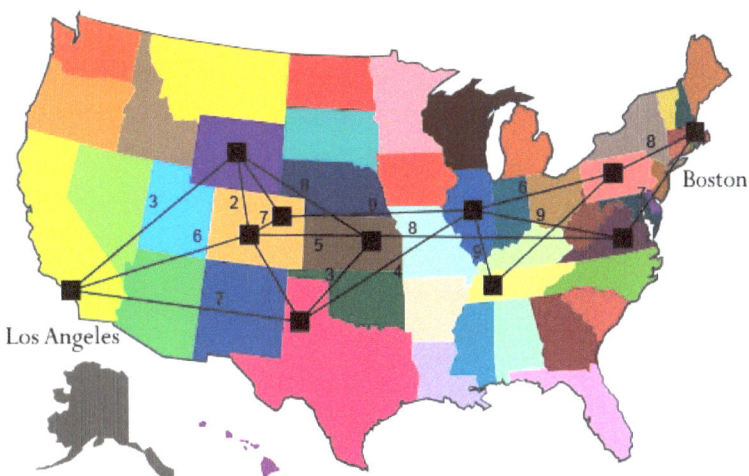

Figure 6.4: A hypothetical phone network connecting Los Angeles and Boston.

The behaviors of ants provide a good solution method to solve complex problems on networks. An ant leaves its nest and randomly chooses a path to look for food.

Figure 6.5: "Knapsack ants". Licensed under CC BY-SA 2.5 via Wikimedia Commons - https://commons.wikimedia.org/wiki/File:Knapsack_ants.svg#/media/File:Knapsack_ants.svg.

If it finds the location of food, it returns to the colony and leaves traces of a chemical on the return path called "pheromone." When another ant finds this chemical, the ant follows the same path to reach the food. More ants that follow this path leave the same chemical markings, making the path more attractive to follow. If the path is long and unattractive, the chemical evaporates and loses its attractiveness as it is not used frequently. If a path is used more frequently, it will be more attractive for an ant as each ant leaves traces of chemical one on top of the other. The most attractive path is the path that is shortest or optimal since it is preferred by many ants that make the chemicals denser and keep it fresh all the time. This algorithm inspired by the behavior of ants is used in practice to solve network problems by mathematically simulating the ants.

Simulated Annealing

Figuratively, simulated annealing is like dropping some bouncing balls over a landscape, and as the balls bounce and lose energy, have them settle down at some local minima. If the balls are allowed to bounce enough times and lose energy slowly enough, some of the balls will eventually fall into the globally lowest locations. Thus simulated annealing is an algorithm that is guaranteed to attain optimality, but only if enough randomness is used in combination with very slow cooling. In practice, there is no guarantee of optimality (and there never is in hard problems).

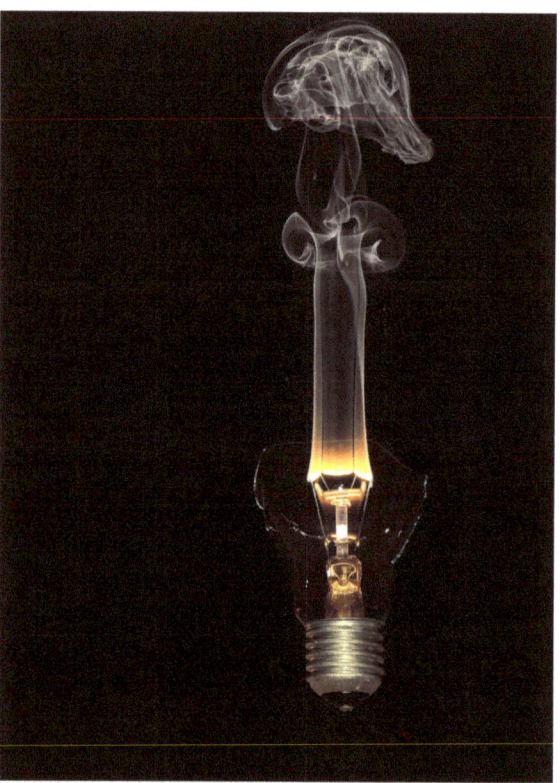

Figure 6.6: Glasses, such as light bulbs, are often annealed. The reason for doing so is that basically an annealed glass product has had its interior stress removed and is therefore much stronger and more stable than an unannealed product.

The metaphor of simulated annealing came from the physical process whereby a solid is slowly cooled so that when eventually its structure is frozen in place, this will happen at the minimum energy configuration. In simpler terms, by cooling slowly, the molecules are able to reach their preferred configurations, thus the solid won't freeze into an awkward, suboptimal configuration, but rather into the ordered, optimal configuration.

Genetic Algorithms

Genetic algorithms are based on natural selection, the biological theory of evolution proposed by Darwin. Roughly speaking, the theory states that traits which increase the likelihood of successful reproduction will become more common in a population. Thus the population evolves towards having more of those traits which are successful.

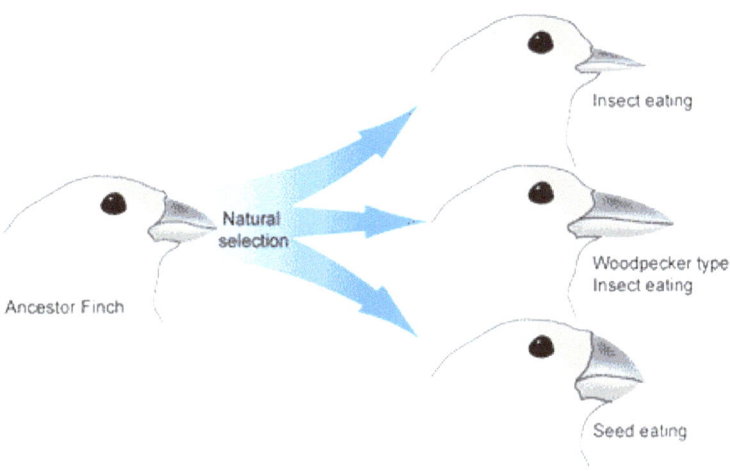

Figure 6.7

Genetic algorithms mimic the process of natural selection mathematically by simulating how genes evolve, mix and prosper in a population.

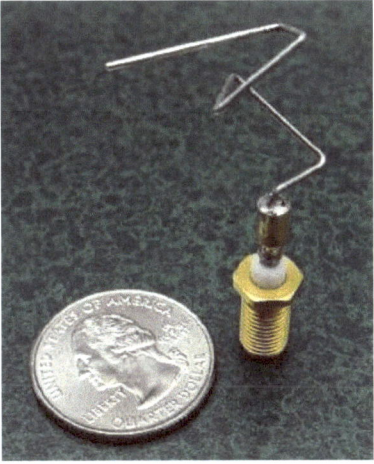

Figure 6.8: The 2006 NASA ST5 spacecraft antenna. This complicated shape was found by an evolutionary computer design program to create the best radiation pattern. It is known as an evolved antenna.

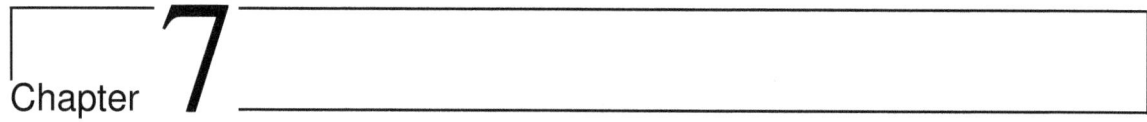

Chapter 7

Population Ecology

The prime-loving Cicadas

Cicadas (Figure 7.1) spend most of their lives as underground nymphs, only emerging after 13 or 17 years. The cicadas emerge all at once. Most likely this is a strategy to reduce losses by satiating their predators. In other words, the predators have their plates so full, they are not able to go after additional cicadas.

It was hypothesized that the emergence period of large prime numbers (13 and 17 years) was a predator avoidance strategy adopted to eliminate the possibility of potential predators receiving periodic population boosts by synchronizing their own generations to divisors of the cicada emergence period.

To illustrate, imagine that the cicadas emerged every 4 years. Some predator might have a litter every 8 years, which would guarantee that the newborn predators always have plenty of chow, and therefore a good head-start in life. Of course, this would be bad for the cicadas. In contrast, if the cicadas were to be born every 7 years and the predator's litter every 5 years, the two events will concur only every 35 years, so that most of the time the litter will not have as good a start in life, and therefore will fare worse. This would of course be great for the cicadas. Having infrequent cycles of emergence of large prime numbers makes this strategy even more effective.

Another viewpoint holds that the prime numbered developmental times represent an adaptation to prevent hybridization between broods with different cycles during a period of heavy selection pressure brought on by isolated and lowered populations during Pleistocene glacial stadia, and that predator satiation is a short term mainte-nance strategy. This complicated hypothesis was subsequently supported through a series of mathematical models, and stands as the most widely accepted explanation of the unusually lengthy and mathematically precise immature period of these insects. Scientists have created models showing that the more years cicadas remained nestled underground, the less likely they would emerge during a killing summer cold spell.

Figure 7.1: Cicadas

Predator-prey equations

The Lotka–Volterra equations, also known as the predator–prey equations, are a pair of first-order, non-linear, differential equations frequently used to describe the dynamics of biological systems in which two species interact, one as a predator and the other as prey, or one as the host and the other as a parasite. The populations change through time according to the pair of equations:

$$\frac{dx}{dt} = \alpha x - \beta xy \tag{7.1}$$

$$\frac{dy}{dt} = \delta xy - \gamma y \tag{7.2}$$

where

- x is the number of prey (for example, rabbits);

- y is the number of some predator (for example, foxes);

- $\frac{dy}{dt}$ and $\frac{dx}{dt}$ represent the growth rates of the two populations over time;

- t represents time; and

- $\alpha, \beta, \gamma, \delta$ are positive real parameters describing the interaction of the two species.

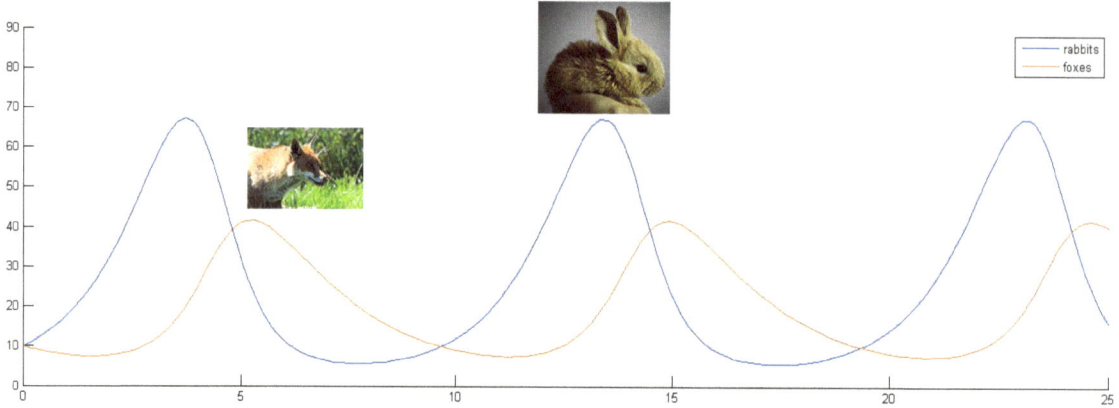

Figure 7.2: Plot of two populations following Lotka-Volterra dynamics.

The Lotka–Volterra models have been used to explain, for example, the dynamics of populations of moose and wolf populations in Isle Royale National Park. Real equations used to model populations are much more complicated, involving additional parameters such as disease, competition and mutualism, but these simple equations serve as prototypes and already exhibit interesting phenomena.

Notice the nice cyclical behavior. At first, prey multiply rapidly. As the prey population soars, predators find more food and begin to grow as well. Eventually the predators eat too many prey and lose numbers to starvation. This lets the prey rebound and the cycle begins again.

Huffaker's mite experiment

In 1958, Carl B. Huffaker did a series of experiments with predatory and herbivorous mite species to investigate predator-prey population dynamics. In these experiments, he created little island universes for the predator and prey mite species. Huffaker was seeking to understand how spatial heterogeneity and the varying dispersal ability of each species affected long-term population dynamics and survival. Contrary to previous experiments on this topic, he found that long-term coexistence was possible under select environmental conditions. This and the following graph provide evidence that the Lotka-Volterra models are relatively accurate under simple conditions.

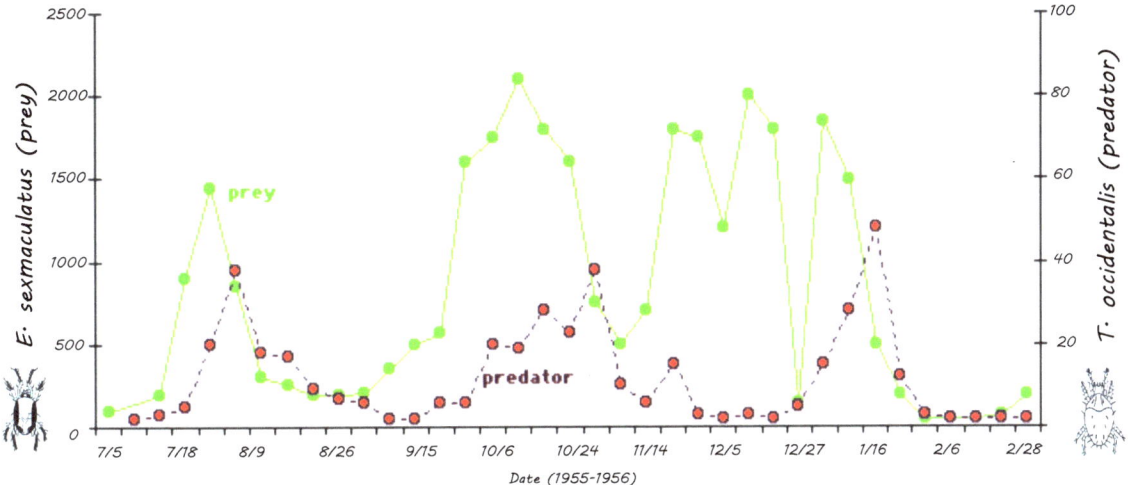

Figure 7.3

It is apparent from the graph that both populations showed cyclical behavior, and that the predator population generally tracked the peaks in the prey population. However, there is some information about this experiment that we need to consider before concluding that the experimental results truly support the predictions made by the Lotka-Volterra model. To achieve the results graphed here, Huffaker added considerable complexity to the environment. Food resources for E. sexmaculatus (the oranges), were spread further apart than in previous experiments, which meant that food resources for T. occidentalis (i.e., E. sexmaculatus) were also spread further apart. Additionally, the oranges were partially isolated with vaseline barriers, but the prey's ability to disperse was facilitated by the presence of upright sticks from which they could ride air currents to other parts of the environment. In other words, predator and prey were not encountering one another randomly in the environment.

Why does the world stay green?

Locusts (Figure 7.4) (and other insects) are famous for their biblical rates of reproduction (literally being featured in the bible). The exponential rate of population growth can be truly frightening. For example, a female housefly, laying a minimum of 600 eggs in her lifetime, would, at the end of a summer of some eight to 10 generations, have 1.9×10^{20} descendants - or roughly 200 million cubic metres of fly. Fortunately, this doesn't happen. Having discussed predator-prey dynamics, it is intuitively clear that the reason is that other predators also grow at exponential rates, and cause a decline in the growth of the prey. The intuitive explanation, however, is likely not correct. It is a point of contention among ecologists, whether the limiting factor in herbivore growth is predators or the poor nutritional value of plants. For example, in [Whi05] it is argued that *the reason herbivores are limited in their population growth is not predators*, but rather that plants (their prey) counteract their predation by using strategies such as poison and making their nutrients harder to access. The limiting factor in the growth of herbivores is the lack of the chemical nitrogen.

Figure 7.4: Locust. CSIRO [CC BY 3.0 (http://creativecommons.org/licenses/by/3.0)], via Wikimedia Commons.

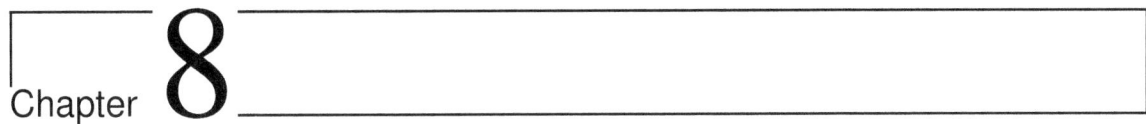

Chapter 8

Human Behavior

> If you ever get close to a human
> and human behaviour
> be ready to get confused

from the song "Human Behavior" by Björk

With the Nobel Memorial Prize in Economic Sciences going to game theorist Jean Tirole in 2014, eleven game-theorists have now won the economics Nobel Prize. *Game theory* is a subfield of applied mathematics that is used by social scientists to quantify and predict behavior. Despite its name, it is not just about games; it is about the strategies that we use every day in our interactions with other people.

To illustrate the type of situations analyzed, consider the *prisoners' dilemma*, a standard example in game theory (Figure 8.1).

Prisoners' dilemma: you and your accomplice have been arrested for a crime. The police place you in separate cells, and you cannot communicate with one another. They then tell you the following: they can convict you both on a lesser charge, but are lacking the evidence to convict you for the big crime you committed. However, if you cooperate, they might be able to work something out with you. In particular, if you rat out your accomplice, and assuming he/she doesn't rat you out, you'll go free and he'll be sentenced to 20 years in prison. If you don't rat him out but he rats on you, he'll go free and you'll be facing 20 years in prison. If you both rat each other out, you'll each get 5 years in prison. If neither of you cooperates with the police, you'll each get 1 year in prison on that lesser charge. The situation is summarized in figure 8.1.

The best outcome for both of you together is if neither one cooperates; however, by not cooperating, you risk facing the worst personal outcome. The best personal outcome occurs if you cooperate, so, no matter what your accomplice chooses, it is best for you personally to cooperate. If he doesn't cooperate, you get out free; if he does cooperate, at least you got only five years rather than 20. *In the prisoner's dilemma, each player pursuing their own self-interest leads both players to be worse off than had they not pursued their own self-interests.* This is a key idea that can be found in many situations in real life.

For example, global climate change can be seen through this theory: if all countries curb CO_2 emissions, then all countries benefit greatly by negating global warming.

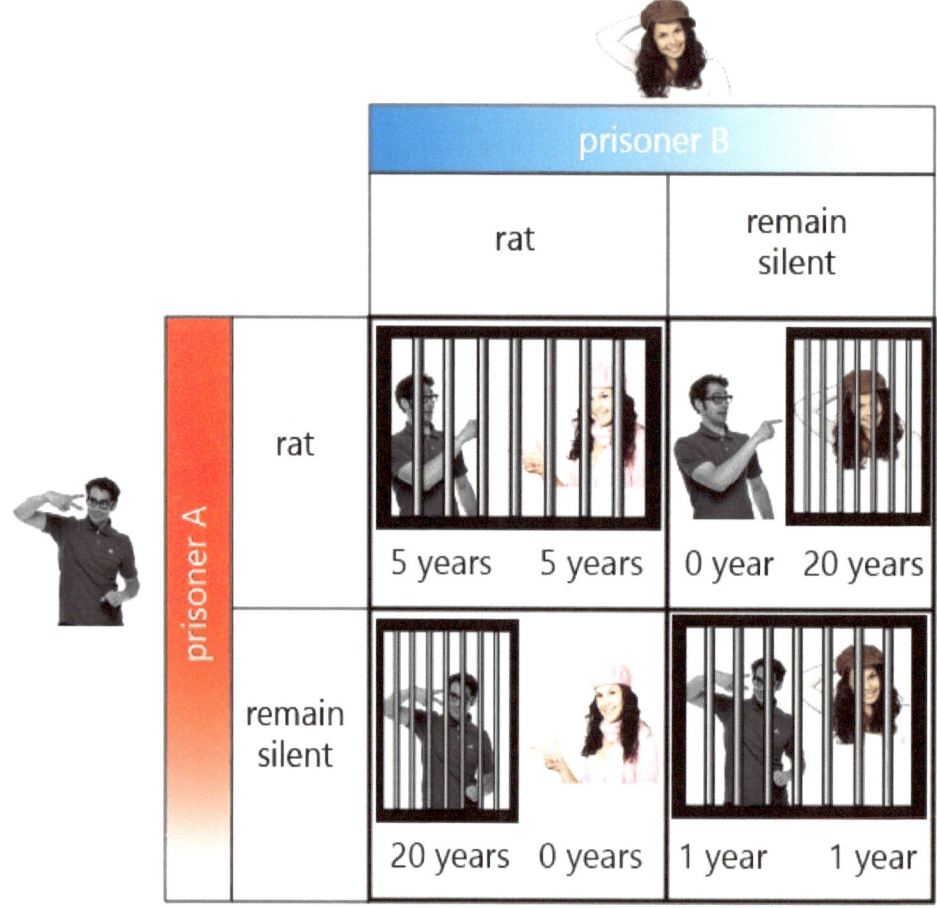

Figure 8.1: Graphical illustration of the prisoners' dilemma. Male prisoner is an adaptation of self portrait by Sebastiaan ter Burg, www.sebastiaanterburg.nl. CC2.0

Consider the situation from the perspective of a single country A. If A cuts down on CO_2 emission and other countries do not, then A has spent valuable resources and got almost nothing from it. However, if A does not cut down on CO_2 emission, then either other countries do and A gets the benefits of not cutting down on CO_2 and better future climate, or all countries do not cut down on emissions and so A still made the best decision by not having cut down on CO_2. This provides a possible explanation for the current impasse concerning climate change.

A second application is to relationships: John Gottman in his research described in "the science of trust" defines good relationships as those where partners know not to enter the

(uncooperative, uncooperative)

cell or at least not to get dynamically stuck there in a loop. Being aware of the prisoners' dilemma uncooperative-uncooperative situation can help avoid it, for example, by communicating the issue and its consequences to your partner.

A few more examples include advertisement, doping in sports, arms races and the treatment of public spaces.

While this will not be discussed in this book, game theory is not only a tool to identify problems, but to generate solutions. Thus scientists often look to game theory

Figure 8.2: Using the case of businessmen arrested for embezzlement as an example of the Prisoner's Dilemma. By Christopher X Jon Jensen (CXJJensen) & Greg Riestenberg - Own work, CC BY-SA 3.0, https://commons.wikimedia.org/w/index.php?curid=19680872

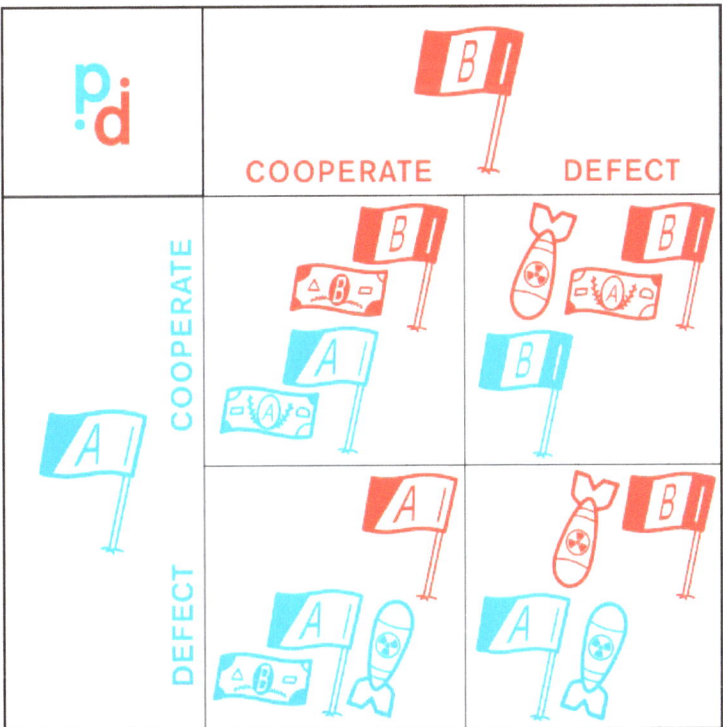

Figure 8.3: Using potential military escalation between two countries as an example of the Prisoner's Dilemma. By Christopher X Jon Jensen (CXJJensen) & Greg Riestenberg - Own work, CC BY-SA 3.0, https://commons.wikimedia.org/w/index.php?curid=19680868

to seek ways to resolve these apparently tragic situations.

Rational Irrationality

Aristotle held the belief that man is a rational animal. A growing body of research suggests otherwise. But whether or not you believe that humans are rational, it is a game theoretic fact that *appearing irrational can be rational*. For example, inmates of mental institutions sometimes "deliberately or instinctively [cultivate] value systems that make them less susceptible to disciplinary threats." [Sch80, p. 17] An inmate could gain some freedom from staff coercion by, for example, becoming self-destructive whenever the mental institution tried to discipline him/her. If a rational person told his boss that he would cut his own vein if demoted, the boss would likely not believe the threat. An inmate in a mental institution, however, could credibly use a threat of self-mutilation to, say, get permission to watch Seinfeld reruns. This certainly seems irrational, since self-mutilation is a high price to pay to get to watch Seinfeld reruns, but if the threat is believable and the staff complies, the inmate benefits.

Revenge, too, is often an irrational behavior. In essence, revenge is harming someone who has harmed you, even if nothing good can come of it. In many documented cases, vengeance is a clearly irrational action. For instance, there are documented cases when people shot their computer! This is obviously only going to make life worse. Yet, this behavior can be explained evolutionarily as follows: the perception that someone will exact vengeance even if it is clear that it will only make things worse for everyone is a great deterrent to provoking this person. It is conceivable that humans have evolved such a trait to put fear in those who try to harm them.

Figure 8.4: A computer shot eight times by a man in Colorado Springs, CO.

Nash Equilibrium

A Nash Equilibrium is a situation in a (game theoretic) game in which all participants have made a decision which they do not regret, assuming they can only change their own action and do not have the option to induce another to change action. For instance, the prisoners' dilemma has two Nash equilibria:

1. (Nash equilibrium) prisoner A and prisoner B rat.

2. (Nash equilibrium) prisoner A and prisoner B both remain silent.

3. (Not Nash equilibrium) one prisoner rats while the other remains silent.

Let's see why these are Nash equilibria, and why they are the only ones.

Suppose that prisoner A and prisoner B both rat. In a Nash equilibrium, the assumption is that one prisoner cannot affect the decision of the other. Thus, prisoner A cannot force prisoner B to remain silent. Hence, for prisoner A to change action to "remain silent" would mean 20 years in jail, much worse than the current 5 years. Consequently, prisoner A has no regrets about his/her decision to rat. The same reasoning applies to prisoner B. Thus situation 1 is a Nash equilibrium.

Similar reasoning applies to situation 2. What happens in the third situation? Suppose that prisoner A remains silent, while prisoner B rats. Now prisoner A certainly regrets his/her decision, meaning that this is not a Nash equilibrium.

Let us put the prisoners' dilemma aside for the time being and consider the familiar game of Rock-Paper-Scissors. This game has no Nash equilibria – no matter which of rock, paper, scissors the players choose, at least one of the players will regret not having chosen something else. Nonetheless, there is an ideal way in which Rock-Paper-Scissors can be played, and this leads us to a different notion of a Nash equilibrium.

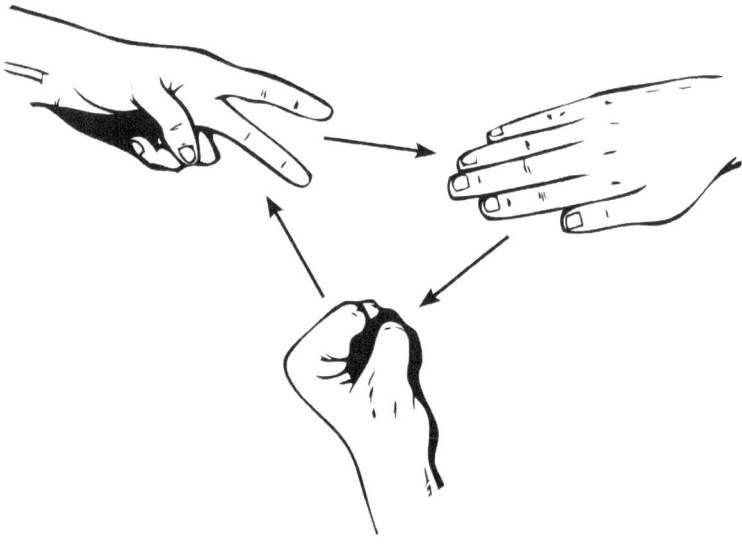

Figure 8.5

If one player has a predictable strategy, the other player is certain to take advantage of it and win. For instance, if one player alternates between rock and paper, then the other will take advantage of it by alternating between paper and scissors. Consequently, a player should randomize his/her strategy by selecting rock, paper, scissors randomly. We call such strategies *mixed*, as there is a mixture of allowed "moves".

Let's suppose that player A has chosen to go with rock 50% of the time, paper 30% of the time, and scissors 20% of the time. Then player B can choose to go with paper 50% of the time, scissors 30% of the time and rock 20% of the time. The result is that player B will win most of the time. Player A will certainly regret choosing this strategy. There is, however, one strategy that neither player will regret taking – a *mixed Nash equilibrium* – and that is to pick each of rock, paper, scissors 1/3 of the time. The lesson here is that by allowing mixed strategies, i.e., strategies which are probabilistic, the game of Rock-Paper-Scissors has a mixed Nash equilibrium – a choice of strategy for each player which is rational and leads to no regrets.

Nash's Existence Theorem is a mathematical theorem that shows that ANY[1] (game theoretic) game has at least one mixed Nash equilibrium.

The proof uses a mathematical technique called a "fixed-point theorem". As the name suggests, in a *fixed-point theorem* there is some operation and this operation leaves some point unchanged. For example, a basketball spun on one finger (Figure 8.6) has two fixed points – the highest and lowest points. Here the operation is the spinning of the ball. Morally, by abstracting and understanding the structure of games, Nash was able to recast the situation so that a fixed-point theorem was applicable, and thus to deduce the existence of a mixed Nash equilibrium.

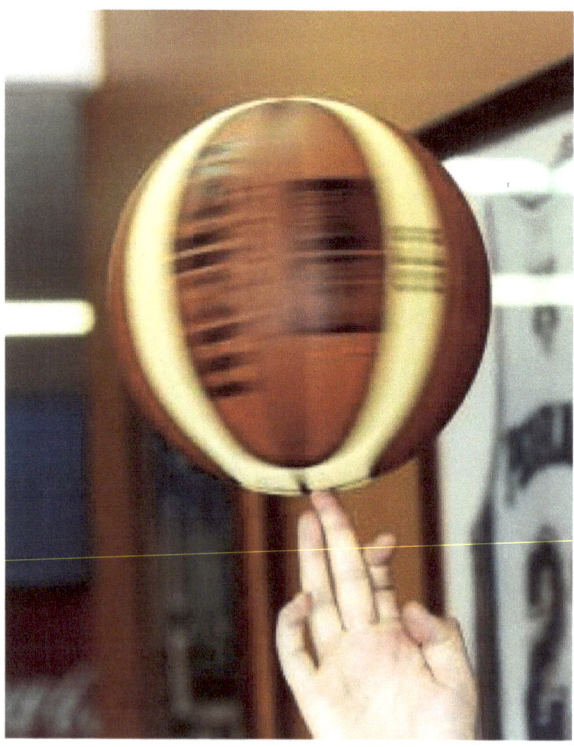

Figure 8.6

---

[1]with a finite number of players in which each player can choose from finitely many pure strategies

Another example of an insight gained from a fixed-point theorem is the following. If you take a map of Paris and throw it somewhere on the ground in Paris, then there is exactly one point on the map that lies on top of the point in Paris which it represents. This is even true if the map is folded.

Choices

Normally, we think: the more choices, the better! The existence of some choices, however, can harm you. This is not only in the case when one's will-power is found lacking, like when you have the option to have ice cream, even though you are on a diet. No. Sometimes having options can make your promises less credible to others. Consequently, by eliminating options you can benefit.

For example, imagine that you are applying to college A. From college A's perspective, admitting you is a risk because it is conceivable that you have also applied to college B and would prefer to go to college B if admitted. Thus A would be reserving a spot for you which it would prefer to provide someone else. In reality, college A is your dream school and you would definitely attend if admitted. You write this in your application to A, but unfortunately, A has no way to be certain that you don't just write "X has always been my dream school" for every college X you apply to. So what to do? Your goal is to convince A that you are truly committed to A. You can do this by eliminating your options and applying through Early Decision (ED) to college A. Early decision plans are binding – a student who is accepted as an ED applicant must attend the college. Thus by eliminating the option to consider other schools, you increase the odds of being accepted into college A, and hence those of getting into your dream school.

Another example is the story of famed explorer and conqueror of the Aztecs, Hernando Cortez. He landed on the shores of Vera Cruz, Mexico in 1519 and wanted his army to conquer the land for Spain. He faced an uphill battle: a terrifying enemy, well-known to sacrifice humans, brutal disease and scarce resources. As he and his army marched inland to do battle, Cortez ordered one of his lieutenants back to the beach with a single instruction: "burn our boats." The effect of this tactic was twofold. First, his enemy now knew that Cortez will fight to the bitter end. Secondly, local tribes considering joining Cortes against the Aztecs – if they previously feared Cortes would abandon them had he encountered early defeat, leaving them alone with the angry Aztec empire – had no longer any doubts about allying with Cortes. Thus even though Cortez now had fewer options (the option to retreat has been eliminated), his odds of success improved.

Evolutionary Game Theory

According to Maynard Smith, in the preface to Evolution and the Theory of Games, "paradoxically, it has turned out that game theory is more readily applied to biology than to the field of economic behaviour for which it was originally designed". That is, even though game theory was conceived to explain human behavior in economics, it has had its greatest successes in explaining biological phenomena. In particular, it has been used to explain many seemingly incongruous phenomena in nature.

As a first example, it was used to explain the evolution and stability of the approximate $1:1$ sex ratios. Here we imagine that a player can choose the sex of its newborn. This is rarely literally the case, but I will explain why it's a reasonable assumption in a moment. If the sex ratios are skewed, with, say, more females than males, then the newborn should be chosen to be male so it has more opportunities to reproduce. This will propagate the strategy of newborn-is-more-likely-to-be-male until the sex ratios become even again. As such, the $1:1$ sex ratios are stable.

Of course most animals neither choose the sex of their offspring, nor calculate the ratio of sexes. Behind the scenes is the following. An ordinary genetic code of an organism is presumably equally likely to give rise to a male or female newborn. Now consider a mutation of this genetic code, for which a male is more likely to be born than a female. We think of this as a "choice" with outcomes. The choice are the odds of male vs. female. The outcome is how likely this genetic code is to propagate. Genes which have a higher likelihood of giving rise to the less highly represented sex will be statistically more likely to be propagated. So even though we use unrealistic language that suggests that an organism chooses actions, this is merely to abstract the fact that behind the scenes, the choices are made probabilistically and then rewarded probabilistically.

Other phenomena, for example, biological altruism – in which an organism appears to act in a way that benefits other organisms and is detrimental to itself, are too explained using game theory. Such behavior is rarely easily explained; the donor sacrifices something directly useful to its survival with the idea that aiding the recipient will pay off eventually. Whether it will and if so, in what manner, can be surprising and unintuitive. Consider for instance insect societies.

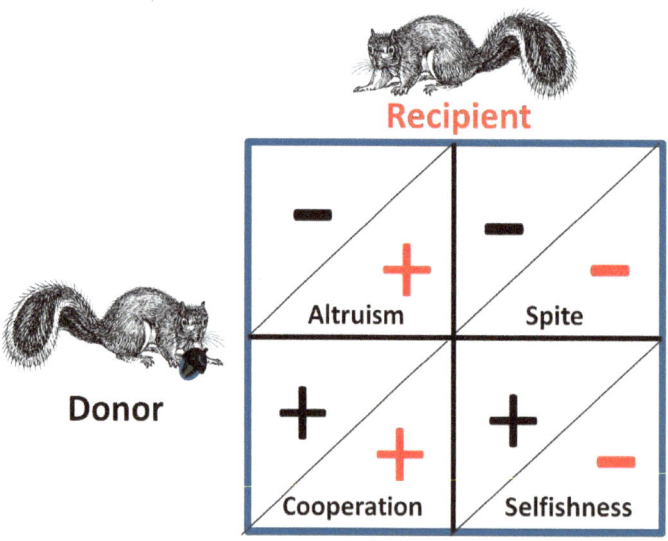

Figure 8.7

Insect societies have always been a source of fascination and inscrutability. Certainly one of the very most inexplicable behaviours of these eusocial insects is the forfeiture of reproductive rights that the workers grant in favour of the queen. In a Darwinian sense, no greater sacrifice can ever exist. Darwin considered this to be one of the most important unexplained phenomena for his theory.

Figure 8.8

To understand this odd behavior, one must note that ants are haplodiploids. This means that males develop from unfertilized eggs and are haploid, whereas the females develop from fertilized eggs and are diploid. Thus a male has no father and cannot have sons, but has a grandfather and can have grandsons. Another consequence is that a female has 3/4 of her genes in common with a sister but only 1/2 of her genes in common with a daughter or son. For comparison, in humans, a person has 1/2 of their genes in common with a brother or sister, and 1/2 in common with a daughter or son. Thus sister ants are more like clones or identical twins than sisters. Consequently, the payoff for female ants is, genetically speaking, higher for having more sisters than children. As a result, only few ants reproduce because this will siphon resources from the queen, and hence, from the production of sisters.

Interestingly, even though it is for the whole ant community's best if only the queen reproduces, a few workers surreptitiously lay eggs of their own, eggs that can grow into reproductive males. Males can fertilize young queens, and if they do, the genetic reward is massive. By diverting shared resources away from the nest, these workers selfishly reduce the fitness of their nestmates. They play the system for their own advantage.

Fortunately for ant communities, ants have a police force to punish criminals. Workers, acting as a force for justice, will attack individuals whose ovaries indicate they might be reproducing. Thus even in ant societies, there is crime and punishment. But it is a much more general feature of nature.

To the unaware, it would seem that crime is "human thing". But actually, just as in ants, occurrences in which an individual takes advantage of the rest of society are very common. In fact, the medical condition known as cancer is an instance in which the cells abandon the collective cause and set out on a (doomed) quest for self-actualization. We see then that corruption and selfishness are commonplace in nature.

In human society, to protect against crime, we have a police force. But what if the police is corrupt? In many countries, police corruption is the norm rather than the exception. As they say, who, then, watches the watchers?

Observation of ants, backed by research in game theory, leads us to a solution to the problem of corruption. The solution is communal justice in the sense that the whole community is ready and able to defend the common good. Such was the case in early human societies, where social norms were enforced by the whole group rather than any specially empowered individuals, e.g., the police. Mathematically, it has been shown that corruption is eliminated and will not reappear stably.

Chapter 9

How the leopard got its spots

What's one thing that we have in our lives that we can depend on? A dog or a cat loving us unconditionally, every day, very faithfully.

Jon Katz

Many phenomena in biology that were believed to be owing to chance or to the action of genes have been revealed to be the consequence of mathematical dynamics. Perhaps the most spectacular example are the patterns on the furs and coats of animals.

Figure 9.1: Leopard. Coat pattern: rosettes

Animals exhibit a remarkable variety of coat patterns – stripes, spots, or even both (Figures 9.1 - 9.7). It would appear that each pattern has its unique and complicated way of forming. Remarkably, a single mathematical mechanism explains the enormous assortment of coat patterns found in mammals.

Figure 9.2:   Jaguar. Coat pattern: rosettes with dot in the center

Figure 9.3:   Zebra. Coat pattern: stripes

The model

Animals start as single cells. Each cell divides into two. Each of those in turn divides into two. This continues on until there are so many cells as to create a full-sized individual. During the early stages, the small ball of cells is completely uniform, looking the same from every direction. But out of this ball develop the dramatic patterns of a zebra, leopard, giraffe, butterfly or angelfish. It is very mysterious that such a uniform ball of cells generates a spatially inhomogeneous static pattern, such as the stripes of a zebra. The famous scientist Alan Turing, credited to be the father of computer science, managed to formulate a series of equations that show very elegantly how the diversity of patterns on animals might be created.

The mathematical model describes the way in which two different chemical products react and are propagated on the skin: one colors the skin, and one inhibits coloration. The inhibiting chemical travels faster than the coloring chemical. Basically, it surrounds

Figure 9.4:   Juvenile tapir. Coat pattern: stripes and spots.

Figure 9.5:   Tiger. Coat pattern: stripes.

the coloring chemical resulting in islands that appear like spots or stripes.

To help visualize this, here is an evocative analogy by the scientist James Murray. There is a field of dry grass in which there is a large number of grasshoppers which generate a lot of moisture by sweating if they get warm. Now suppose that fire breaks out in the field at several random points and a flame front starts to propagate from each. We can think of the grasshopper as an inhibitor and the fire as the coloring chemical. If there were no moisture to quench the flames the fires would simply spread over the whole field, which would result in a uniform charred area. However when the grasshoppers get warm enough they generate enough moisture to dampen the grass so that when the flames reach such pre-moistened areas the grass will not burn. Thus, the fires start to spread; when the grasshoppers ahead of the flame fronts feel it coming they move quickly ahead of it. The grasshoppers then sweat profusely and generate enough moisture to prevent the fires spreading into the moistened area. By this way, the charred areas are restricted to domains which depend on the speed of the fire, of the grasshoppers, and shape of the forest.

Predictions

What is remarkable is that the equations show that the different patterns of coat depend only on the size and form of the region where they are developed. Stated in another way, the same basic equation explains all of the patterns. This does not contradict the observation that the tiger and the leopard have different patterns given that their bodies are similar because the formation of the patterns would not be produced at the same moment during the growth of the embryo. In the first instance, the embryo would be still small and in the other, it would be at a much bigger stage. The equations show that no pattern is formed if the embryo is very small, that a striped pattern is formed if the embryo is a little bigger, a spotted pattern if it is bigger yet, and no pattern at all if it is too big. This is why the mouse and the elephant do not have a pattern.

From the theory, one can actually obtain "biological theorems" of the following form.

Theorem: Snakes always have striped (ring) patterns.

Figure 9.6: Snake. Coat pattern: striped ring pattern.

In particular, snakes do not have spotted patterns. This is because we think of snakes as being very skinny, almost 1-dimensional. If you imagine a wave in a thin vessel, it will always travel in the long direction. The model for coat patterns is not much different.

Theorem: A striped animal cannot have a spotted tail.

That is, there are animals with striped bodies and striped tails (tiger, Figure 9.5), spotty bodies and spotty tails (leopard, Figure 9.1), and spotty bodies and striped tails (cheetah, genet, Figure 9.7), but never a striped body with a spotted tail. The reason is that the body is always wider than the tail. The same mechanism is responsible for the patterns on both body and tail. Then if the body is striped, and the parameters are similar for tail and body, then the tail must also be striped since the narrower geometry makes it easier to produce strips.

Figure 9.7: Genet. Coat pattern: spotted body, striped tail. By Guérin Nicolas (messages) (Own work) [GFDL (http://www.gnu.org/copyleft/fdl.html) or CC BY-SA 3.0 (http://creativecommons.org/licenses/by-sa/3.0)], via Wikimedia Commons

Seashells

Old style thermostats made use of a bimetallic strip (Figure 9.8). One side, consisting of one type of metal, expanded faster due to temperature increase than the other side which consisted of another metal. As a result, the strip would bend, and at a certain temperature break an electrical connection.

A similar principle explains the patterns and shapes of seashells. Seashells consist of calcified material. The animals can increase the size of their shells only by accretion of new material at the margin, at the growing edge of the shell (but not the other edge). Thus there is an asymmetric expansion at the margin, just as in the bimetallic strip.

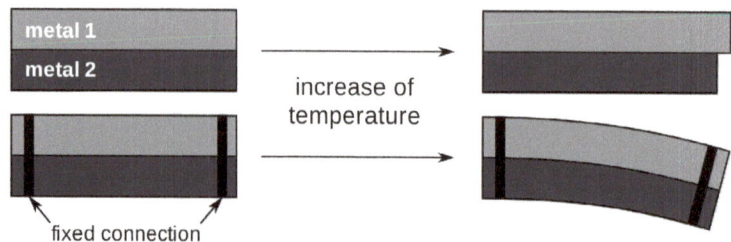

Figure 9.8: A bimetallic strip. By Patrick87 - Translation of Bimetallstreifen.svg, CC BY-SA 3.0, https://commons.wikimedia.org/w/index.php?curid=25482530.

Hence the seashell continues to bend in three-dimensions, producing the spiral shapes they are so famous for.

Figure 9.9: A photograph of the snail Oliva porphyria. By Hectonichus - Own work, CC BY-SA 3.0, https://commons.wikimedia.org/w/index.php?curid=42823771.

Most decorations of shells result from the incorporation of pigments during this growth process. Once made, as the rule, the patterns remain unchanged. The patterns are therefore historical records of what happens at the growing edge, i.e., they are a time record of the pattern formation process.

Afterword

This book was written with the intention of illustrating how mathematics can be enjoyable in its own right, like a well-composed sonata or a peaceful morning of bird watching. In a way, the task is simple. After all, just look at your garden to find mathematical beauty and elegance. It is there, independent of who we are and what we do, and had I been religious, I would point to this as proof that a divine being loves us, unconditionally. Of course, simple tasks, as often turns out, are not easy tasks, and it was quite the challenge to find the golden mean between too technical and too elementary.

In writing this book, I was driven by love, but also by frustration. Many times I have heard everyday folks express disliking mathematics. In my eyes, this stems not from understanding mathematics and finding it distasteful, but from associating mathematics with the educational system. Thus, "I hate maths!" is actually code for "I hate the education system!". And the education system really does sin in the math department. Two important things are generally lacking in mathematics education:

1. Real and convincing illustrations that mathematics is interesting

2. Proper explanation of how mathematics is applied

Addressing the first issue was the goal of this book. In my everyday life, I found that people enjoyed hearing some of the mathematical tales I have compiled in this book, and were generally unaware that elegant math is all around is, waiting to be discovered.

The second issue is the lack of a convincing answer to the often-heard question, "how am I ever going to use this?". Word problems in school give an overwhelming impression that math is completely useless and divorced from reality. Problems like

"Emily loves to fish during summer vacation. The first day, Emily catches 3 fish. The second day, Emily catches 6 fish. The third day, Emily catches 9 fish. The fourth day, Emily catches 12 fish. If this pattern continues, how many fish does Emily catch on the eleventh day? Show all of your mathematical thinking."

But in reality, it is incredibly easy to answer this question, or at least the question "how is math useful?". Our whole world runs on math: statistics, internet security, circuits, airplanes, GPS, thermostats, finance,... A convincing explanation of how math is used in such areas is a missing component in education. I hope to address this issue in a future piece.

Bibliography

[Bar52] C. S. Barret. *Structure of Metals*. McGraw-Hill, 2nd edition, 1952.

[FT07]   Dmitry Fuchs and Serge Tabachnikov. *Mathematical omnibus*. American Mathematical Society, Providence, RI, 2007. Thirty lectures on classic mathematics.

[Sch80] T.C. Schelling. *The Strategy of Conflict*. Harvard University Press, 1980.

[Whi05] Thomas C. R. White. *Why Does the World Stay Green?: Nutrition and Survival of Plant-eaters*. Csiro Publishing, 2005.

www.ingramcontent.com/pod-product-compliance
Lightning Source LLC
Chambersburg PA
CBHW050735180526
45159CB00003B/1232